# Civil Engineering Practice in the Twenty-First Century

*Knowledge and Skills for Design and Management*

Neil S. Grigg
Marvin E. Criswell
Darrell G. Fontane
Thomas J. Siller

American Society of Civil Engineers
1801 Alexander Bell Drive
Reston, Virginia 20191-4400

Abstract: In addition to designing, building, and managing the constructed environment, civil engineers have important environmental management responsibilities. To succeed in these tasks, civil engineers must possess both management skills and technical tools. The Accreditation Board for Engineering and Technology, studies of the civil engineering work force, and engineering practitioners confirm the need for such skills and tools. The 12 chapters that comprise this book will assist students in developing those skills and prepare practitioners for the civil engineering challenges of the twenty-first century. Chapters discuss such topics as civil engineering heritage and future, consequences of civil engineering, work and careers in civil engineering, engineering design and the infrastructure life cycle, management, critical thinking, communications, government, finance and economics, law, and professional practice and ethics.

Cover photo of highway construction is courtesy Colorado Department of Transportation, 1999; photo by Gregg Gargan.

Library of Congress Cataloging-in-Publication Data

Civil engineering practice in the twenty-first century / Neil S. Grigg ... [et al.].
    p. cm.
Includes bibliographical references and index.
ISBN 0-7844-0526-3
1. Civil engineering.   I. Grigg, Neil S.  II. American Society of Civil Engineers.
TA145 .C59  2001
624—dc21                                                                                                         2001018217

The material presented in this publication has been prepared in accordance with generally recognized engineering principles and practices, and is for general information only. This information should not be used without first securing competent advice with respect to its suitability for any general or specific application.

    The contents of this publication are not intended to be and should not be construed to be a standard of the American Society of Civil Engineers (ASCE) and are not intended for use as a reference in purchase of specifications, contracts, regulations, statutes, or any other legal document.

    No reference made in this publication to any specific method, product, process, or service constitutes or implies an endorsement, recommendation, or warranty thereof by ASCE.

    ASCE makes no representation or warranty of any kind, whether express or implied, concerning the accuracy, completeness, suitability, or utility of any information, apparatus, product, or process discussed in this publication, and assumes no liability therefore.

    Anyone utilizing this information assumes all liability arising from such use, including but not limited to infringement of any patent or patents.

ASCE and American Society of Civil Engineers—Registered in U.S. Patent and Trademark Office.

Photocopies: Authorization to photocopy material for internal or personal use under circumstances not falling within the fair use provisions of the Copyright Act is granted by ASCE to libraries and other users registered with the Copyright Clearance Center (CCC) Transactional Reporting Service, provided that the base fee of $8.00 per chapter plus $.50 per page is paid directly to CCC, 222 Rosewood Drive, Danvers, MA 01923. The identification for ASCE Books is 0-7844-0526-3/01/$8.00 + $.50 per page. Requests for special permission or bulk copying should be addressed to Permissions & Copyright Department, ASCE.

Copyright © 2001 by the American Society of Civil Engineers.
All Rights Reserved.
Library of Congress Catalog Card No: 2001018217
ISBN 0-7844-0526-3
Manufactured in the United States of America

# Contents

|    | Foreword                                              | v    |
|----|-------------------------------------------------------|------|
| 1  | Civil Engineering in the Twenty-First Century         | 1    |
| 2  | Civil Engineering: History, Heritage, and Future      | 13   |
| 3  | Consequences of Civil Engineering                     | 45   |
| 4  | Work and Careers of Civil Engineers                   | 61   |
| 5  | Engineering Design and the Infrastructure Life Cycle  | 83   |
| 6  | Management for Civil Engineers                        | 109  |
| 7  | Critical Thinking                                     | 139  |
| 8  | Communication for Civil Engineers                     | 149  |
| 9  | Civil Engineers and Government                        | 163  |
| 10 | Economics and Finance for Civil Engineers             | 187  |
| 11 | Law for Civil Engineers                               | 217  |
| 12 | Professional Practice and Ethics                      | 233  |
|    | Appendix: ASCE Code of Ethics                         | 243  |
|    | References                                            | 249  |
|    | Index                                                 | 257  |

# Foreword

In contributions to society, civil engineers are hard to beat. They plan, design, construct, operate, maintain, and rebuild infrastructure and environmental systems that are critical to the survival of the human race and vital ecologic systems. Yet at the beginning of the twenty-first century, civil engineers find that their technical skills must be supplemented more than ever by other skills such as critical thinking, communications, and management.

This point has been driven home for us by civil engineers with years of experience as well as by graduates just entering the work force. What we are hearing is echoed in other places as well, including work force studies, university core curriculum advances, visiting practitioners, and the Accreditation Board for Engineering and Technology. Other professionals who seek to stay current are experiencing the same pressures.

In responding to this clear need for skills that extend beyond the technical arena, we have found that the most critical to practice today are thinking, management, and communications skills. This places us on the horns of a dilemma—does each student take a single course in a subject such as critical thinking, or should critical thinking be infused into all courses in the curriculum? We argue that the latter is the only feasible route, and we have introduced an integrated curriculum to bring the critical skill areas into learning about civil engineering.

This book presents our approach to the integration of essential skills into the civil engineering curriculum. It does not include many of the technical areas but can be used alongside technical textbooks to guide students and practicing civil engineers in updating and broadening their capabilities with the skills they need to succeed in the twenty-first century.

This text derived from courses given in the Department of Civil Engineering at Colorado State University. We would like to thank Ms. Bernadette Shepard, who helped with administrative work, and the department faculty and students who contributed ideas and materials. Mike Kuyper, one of our students in 1998–1999, contributed the art work for Figure 1-1.

Neil S. Grigg
Marvin E. Criswell
Darrell G. Fontane
Thomas J. Siller
*Colorado State University*
*Fort Collins, Colorado*

# 1 Civil Engineering in the Twenty-First Century

## Introduction

Although civil engineering will continue to be a vibrant and rewarding profession into the twenty-first century, the type of work associated with the profession will change dramatically. Rapid technologic and population growth as well as increasing environmental concerns will shape both the profession and individual careers. How civil engineers respond and adapt to these and other challenges will determine the vitality of the profession as a whole as well as individual success.

All professions add to the mix of human effort, enabling us to survive and progress. Likewise, every profession will be subject to change in the future, and each group must undergo self-examination to adapt successfully. What will civil engineering be like in 50 years? Technology has progressed so rapidly that engineering practice is much different today than it was even 2 decades ago. Social change has also been dramatic, and changing mixes of government and privatization influence our work. There is much to do that requires adjustments in how we are educated. Although such basic skills such as graphics, computation, and analysis will still be required, civil engineers will be expected to do much more in the future.

Civil engineering educators receive a great deal of advice from alumni and practitioners. Some of the most often repeated sentiments include the following: "Keep teaching the basics, but send us graduates who can communicate better, who understand the business world, and who know something about finance," and "Don't neglect technical subjects, but the most important thing is that your graduates think clearly and exercise good decision-making skills." Such comments suggest that there is much more to civil engineering practice than technical knowledge. Accrediting agencies also know this and have developed a certification process for engineering programs to validate a broad education.

This book was written in response to the many suggestions we have received about civil engineering education. It outlines how the profession is changing and how students can prepare for such change. We hope it will be a helpful companion for civil engineering students and practitioners who navigate the shoals of constantly changing opportunities and challenges. This book results from practitioners' answers to the question of what should civil engineering students be taught. The answer is that civil engineering students should be taught many subjects that go beyond the standard engineering curriculum.

It is impossible for a 4-year engineering program to include courses covering every subject students may need or want. Our approach is to integrate these topics into other courses during the B.S. program. This method provides preparation for life, which requires dealing with many situations simultaneously. Civil engineers have two main roles: building and managing infrastructure and sustaining environmental resources. Carving out meaningful careers in these arenas while also adapting to change will be an exciting and potentially rewarding challenge.

## Civil Engineering in the Twenty-First Century

What will civil engineering work be like in the twenty-first century? Civil engineers can look forward to being involved in society's most important problems. Figure 1-1 shows the six primary infrastructure systems that impact civil engineering work as well as the impact of civil engineering on society and the natural environment. These infrastructure systems are discussed in depth throughout this text.

Some forecasts about future work, such as those listed in the 1987 publication *Workforce 2000*, accurately predicted today's conditions (Johnston and Packer 1987): (1) The economy would be strong; (2) manufacturing would shrink as a percentage of the economy but would not wither away; (3) the work force would grow slowly and consist of older people, more women, and more minorities; and (4) jobs in service industries would require new skill levels.

*Workforce 2020*, published in 1997, predicts that the pace of technologic change will grow, the rest of the world will matter more to the United States, America will get older, and the labor force will continue its ethnic diversification (Judy and D'Amico 1997). These predictions are clear and appear to be accurate. The impact of these trends on civil engi-

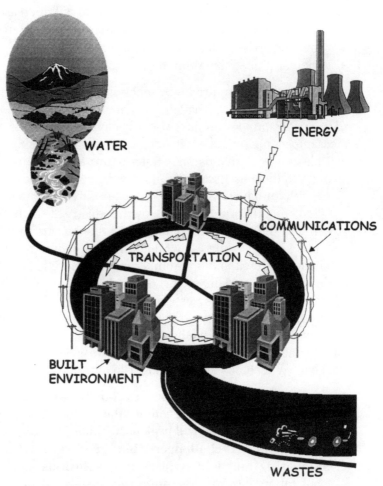

**Figure 1-1**  Civil infrastructure systems.

neering careers, including how to adapt new technologies to infrastructure systems, how to adapt to the global environment, and how to respond to a more diverse working environment, are discussed throughout this book.

Civil engineers must adapt all available information about professional trends and career advice, including trends published by the Career Center at the University of Waterloo, Canada (University of Waterloo 1997):

- The workplace is changeable.
- The job market is dynamic and changeable.
- Positions will be less stable, and lifetime employment will end.
- Organizations do not take responsibility for employee development.
- Individuals must take responsibility for their careers.
- Downsizing and delayering will reduce promotion opportunities.

- Outsourcing and giving work to suppliers will increase.
- Joint ventures will create opportunities for entrepreneurs.
- Small businesses must compete internationally.
- There is a need for computer literacy.

Career trends are driven by technologic, economic, and social developments. The most dramatic trends are in the information and communications technology arenas. Also, the global stage of business will have a great impact on engineers. Population growth and rising living standards will place new demands on infrastructure and the environment. Many changes in business practices and government lie ahead. Some of these, along with implications for civil engineers, are presented in Table 1-1.

Predicting the future was a growth industry at the beginning of 2000. To usher in the new millennium, the Massachusetts Institute of Technology (MIT) organized the New Millennium Colloquium, which deliberated on changes in civil and environmental engineering. Wayne Clough, civil engineer and president of the Georgia Institute of Technology (which graduates the nation's largest group of engineers), had a big influence on the group. Clough believes that the combination of a growing world population and the human tendency to delay infrastructure and environmental improvements indicates that civil engineers will be essential in the future. His ideology includes the following (Clough 2000):

- **Large population increases are on the way.** For example, during the latter part of the next century the United Nations anticipates that the world population will peak at 10 billion, up from today's 6 billion.
- **Basic societal challenges that require civil engineers are increasing and will require new solutions.** Society has a blind spot for its basic needs, preferring instead to focus on glamorous developments. Problems that result from this thinking involve housing the population, addressing decaying urban infrastructure, maintaining the environment, dealing with natural disasters and climate changes, and transporting people and goods. Sustainability of the environment and society will be paramount issues.
- **Addressing the challenges will require civil engineers to apply new technologies.** This includes biotechnology, materials and nanotechnology, electronic commerce, advanced communications, and information technology. Research and development monies are in short supply in our field because they are being shifted to these technologic areas. The technology that drives solutions is unlikely to start with civil engineers; however, a large part of our job is to adapt to technologies.
- **Everyone (including civil engineers) must adapt new management strategies and respond to taking business online and streamlining the construction process.** Today's business environment involves rapid change, and engineers will be able to utilize

**Table 1-1** Future government and business trends and implications for civil engineering.

| Future Trends | Civil Engineering Implications |
|---|---|
| *Globalization*<br>— Global economy<br>— Instant communication<br>— Cultural integration<br>— Jobs and immigration | — There will be dramatic changes in civil engineering careers, businesses, and markets. |
| *Population and development*<br>— Population increases<br>— Sustainable development<br>— Consumption demand | — There will be more work in the areas of infrastructure improvement, environmental management, and protection against disaster loss.<br>— Clients may have difficulty paying for work in some cases. |
| *Technology and knowledge*<br>— Technology acceleration<br>— Knowledge advancement<br>— Knowledge diffusion<br>— Green technologies | — Civil engineers must adapt to new technologies and work modes. |
| *Government*<br>— Triumph of democracy<br>— Capitalism and socialism<br>— Privatization<br>— Partnerships | — Organization of the civil engineering industry will change to facilitate adaptation to increasing marketplace competition. |
| *Social shifts*<br>— Equity and poverty<br>— Multiculturalism<br>— Internationalization<br>— Public health<br>— Social tensions | — Civil engineers have the opportunity to respond to social needs, if they can develop designs and management plans that meet human needs, and simultaneously satisfy both public and private clients. |
| *Business*<br>— Change rate acceleration<br>— Internationalization<br>— E-commerce and business-to-business<br>— Mergers and acquisitions<br>— Vertical integration<br>— Life cycle products<br>— Green business | — Businesses and government will change rapidly.<br>— Civil engineers must be flexible and continue to learn about new business trends and possibilities. |
| *Work*<br>— Teams<br>— New and different workers<br>— End of 8-hour workday<br>— Workplace portability (ability to work anywhere) | — The workplace will change dramatically.<br>— Civil engineers and other professionals will not be entitled to any particular market but must continue to demonstrate that their skills prepare them to deliver superior products and services. |

Internet–based technologies and management innovations to improve the construction process, while serving clients better.
- **A new kind of civil engineer is required.** In the past, civil engineering was considered a primarily technical field, and the technical focus will continue, of course. The new kind of civil engineer will require additional skills, however, that enable him or her to communicate better than in the past and succeed in a faster-paced business and political environment.
- **The civil engineer of the next millennium must be educated differently than in the past.** Civil engineering must attract its share of the best and the brightest amid a rapidly changing and diversifying work force.

Given the trends outlined above, what must civil engineers do to succeed in the twenty-first century? We can take a lesson from Clough's point that basic challenges that require the expertise of civil engineers are increasing in number and importance and will require new solutions but that society has a blind spot for these challenges. Civil engineers will continue to serve in critical societal arenas, but the spotlight may prove elusive.

Civil engineers work on infrastructure and environmental problems in both public and private endeavors, including consulting firms; local, state, and federal government; construction; environmental organizations; and other organizations. Civil engineers must assume such diverse roles as project planner and advocate, regulator, analyst and designer, and builder. Civil engineers may specialize in structures, hydraulics, environmental engineering, transportation, geotechnical, or other fields but will likely be required to work as an entry-level designer, construction manager, or chief executive officer.

Civil engineering work is different from that of other engineering disciplines in that it involves more public sector spending and regulation; it involves private practice more and attracts more interest in professionalism; it has more influence on the construction and infrastructure industries and on environmental regulation; it has a larger social component; and it is more stable than other engineering disciplines and the size of the occupational group is not increasing rapidly. An artist's perception of the engineer's public work is shown in Figure 1-2.

## Preparing the Civil Engineer

Educators, universities, and the American Society of Civil Engineers (ASCE) have studied how engineers should prepare for the challenges of the twenty-first century. These three sources provide civil engineers with guidance for career preparation.

Civil Engineering in the Twenty-First Century ♦ 7

**Figure 1-2**  An artist's perception of Public Works Week, a period of national focus on infrastructure. (Courtesy American Public Works Association)

The Accreditation Board of Engineering and Technology (ABET) educational requirements (ABET 1999), which are directed at all engineers, include the following:

- Mathematics, science, and engineering
- Design and conduct of experiments
- Analysis and interpretation of data
- Design of systems, components, and processes
- Functioning on multidisciplinary teams
- Identifying, formulating, and solving engineering problems

- Professional and ethical responsibility
- Communication
- Global and societal impact
- Lifelong learning
- Contemporary issues
- Use of techniques, skills, and tools
- Professional component (integrated experience)

Institutions of higher education normally have core curricula that lay out the minimum requirements for all students who look ahead to careers, family responsibilities, and other challenges. Categories of such curricula are similar across the country and include subjects such as this list from Colorado State University:

- A freshman seminar to provide an integrative experience for first-year students
- Written and oral communications to establish functional skills in communication, possibly including a second language
- Mathematics and basic computing skills
- Logical/critical thinking to lay a foundation for analysis and problem solving
- Basic knowledge in science areas
- Arts/humanities to provide at least a basic appreciation of these areas
- Social/behavioral sciences to provide a foundation of knowledge in these areas
- Historical perspectives to provide a sense of the past
- Awareness of current context and events
- Knowledge of how society and government evolved in the United States
- Appreciation of personal health and wellness
- Integrating competencies: writing, speaking, and problem solving to demonstrate the ability to apply communication and thinking skills
- Demonstrating the ability to apply knowledge to solve problems
- Integration of knowledge and skills

Engineers will follow similar curricula but will require further study in some areas (e.g., computation and problem solving) and more specific knowledge in others (e.g., a sense of the heritage of their own profession).

## Civil Engineering Education

Career success stems from many factors, including background, education, experience, motivation, and luck. Education and experience are aspects of success that can be controlled: Much can be learned and experienced during 4 years of undergraduate education that can help to lay a foundation for lifelong learning.

Engineers should acquire a set of knowledge, skills, and abilities (KSAs). These KSAs should be rooted in the skills needed by all professionals, managers, and citizen leaders who must think critically, communicate with others, deal with such public issues as the operation of government and public finance, and function at high levels in today's complex society.

Technical skills are the foundation of the civil engineer's preparation for professional practice, but many practitioners report that management, communication, and problem-solving skills are also critical. A combination of these skills is required for success in any profession, and civil engineers must continue to learn in order to experience a successful and satisfying career.

Although required skills are similar for many professions, civil engineers apply those skills in different ways. For example, civil and mechanical engineers have many skills in common, but they work in different industries. Likewise, civil engineers and lawyers must think critically, although with different training and in different settings. Working with government agencies means different things to civil engineers, lawyers, and doctors. Learning essential skills carries with it a requirement to apply those skills appropriately and in context.

Some rapidly advancing areas, such as computing, almost drive themselves because those who refuse to learn will simply be left behind. Traditional lines of civil engineering work (e.g., fluid mechanics, materials, structures, geotechnical engineering) also advance, although not as rapidly as computing and telecommunications.

New graduates, practicing engineers, and industry publications have all noted that civil engineers need both basic and continuing education in engineering, computing, and other essential skill areas. These areas are focused on civil engineering practice, which has technical and management tracks. Figure 1-3 illustrates how civil engineering practice is guided by input from different areas of education and preparation. It is important to understand the basics of each of these categories of knowledge.

## Basic Technical Subjects

Basic technical subjects, such as mathematics and physics, form the entry portal for engineering, but they only begin to shape the engineer's capabilities. For example, some engineers may not need to use differential equations in their working life, but learning that skill enables them to handle quantitative analysis and provides the rigorous training in problem solving that is required by all engineers.

## Applied Technical Subjects

Applied subjects form the technical core of civil engineering. The major subjects, sorted by enrollment in ASCE technical divisions (as of

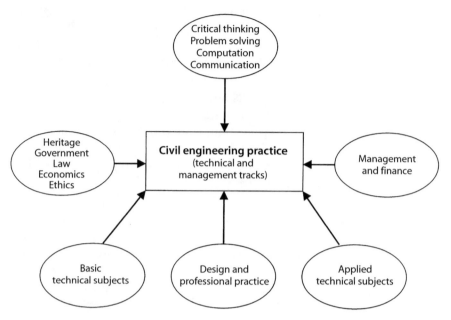

**Figure 1-3**   Areas of input for civil engineering practice.

1997) are water resources, transportation, construction, structures, environmental engineering, and geotechnical engineering. These areas create the need for one or more applied technical courses in engineering. For example, learning about water resources normally requires at least two courses (hydraulics and hydrology). Transportation involves several areas, beginning with highway engineering. Construction attracts a large enrollment of civil engineers in ASCE but is not necessarily an area for a single course; this topic might be spread across several courses. Structures involve several course areas, beginning with analysis and extending to steel, concrete, and other materials. Finally, both environmental and geotechnical engineering require at least one introductory course each.

In addition, other cross-cutting (niche) specialties (computer practices, urban planning, geomatics, waterways and ocean engineering, pipelines, forensics, materials, engineering mechanics, energy, earthquake engineering, architectural engineering, and cold regions engineering) attract significant enrollment. Each of these subjects should be covered to some extent in a basic education but would not normally contribute specific courses to the undergraduate curriculum.

## *Design*

Design is a skill area that is difficult to pin down because it involves a mixture of skills ranging from graphics to creative thinking to experience-based knowledge of ways to approach different situations. Design is central to civil engineering work. Much of the civil infrastructure in the United States has been built since World War II, and it will soon require rehabilitation and replacement. This, combined with new growth, demands increasing amounts of design work.

The design process includes the infrastructure life cycle, from recognition of a need to functioning of the constructed system. Viewing the infrastructure life cycle, we identify processes of planning, conceptual design, preliminary design, final design, construction, operation, maintenance, and rehabilitation/replacement or decommissioning/demolition. Design is discussed in more detail in Chapter 5; however, it should be noted that the subject is too broad to be covered comprehensively in just one chapter.

## *Professional Practice*

Professional practice is the domain of knowledge that best expresses the context aspect of civil engineering. Knowing the norms of practice, forms of project delivery, ethics, and professional situations requires actual work experience. The various aspects of professional practice are discussed throughout this book.

## *Critical Thinking, Problem Solving, Computation, and Communication*

All professionals must be skilled in critical thinking, problem solving, computation, and communication. It is arguable whether civil engineers require more skill in these areas than do groups with similar responsibilities. This is discussed in more detail in Chapters 6 and 7.

## *Heritage, Government, Law, Economics, and Ethics*

The social studies component of civil engineering includes heritage, government, law, economics, and ethics. These subjects are taught in kindergarten through 12$^{th}$ grades and are part of the general education provided in universities. However, knowledge in the field of civil engineering must extend to the context of modern pressures and realities.

Civil engineers must be skilled in law and conflict resolution because our industries (infrastructure and environment) involve many conflicts. Civil engineers also work to protect the environment and are vitally involved in sustainable development (focused on habitat, clean water and air, proper land use, quality of life, reuse and recycling, and stewardship of resources) and in responding to global change. Sustainable development is discussed in more detail in Chapter 3. This book reviews the work of early engineers as well as the consequences of modern civil engineering.

One way in which civil engineering work has changed is in the increased regulation of our society, particularly with respect to the environment. Civil engineers are engaged in much of this regulatory work, preparing environmental impact assessments; studying total maximum daily loads of water contaminants; and dealing with other aspects of clean water, air quality, land use management, waste management, and habitat. In addition, construction safety is closely regulated, as are building codes and many other aspects of civil engineering. Chapters 9 and 11 can provide a perspective on how regulatory work is approached.

### *Management and Finance*

The degree to which civil engineers are involved in management and finance is not always recognized, but civil engineering educators continually hear about this from graduates and advisory groups. Many civil engineers who do find themselves in management situations find that they are unprepared. This often occurs much sooner than they would expect. Management requires problem-solving and critical-thinking skills. As a result of their close relationship with construction, civil engineers are often involved in project management (see Chapters 5 and 6).

Civil engineers use many tools, including engineering analysis and design, quantitative descriptors and graphs, and computer tools (including computer-aided design and geographic information systems). More advanced management-oriented tools include systems thinking, process mapping and models, case studies, and strategic planning. Teamwork remains a critical ingredient.

## Lifelong Learning

We hope you enjoy this book. It has been written to help launch your lifelong learning about the civil engineering profession. The material presented is designed help young engineers learn about the profession and to give tips to practicing engineers about subjects they may not have considered. Civil engineers are members of an old and proud profession and have much to contribute to society.

# 2 Civil Engineering: History, Heritage, and Future

## Introduction

This chapter describes the rich history and heritage of civil engineering and illustrates how civil engineers have made history, just as generals and politicians have. The history and heritage of civil engineering are vast but can be sampled by tracing how civil engineering structures and services have evolved during certain epochs. There are many themes to the story, and this chapter examines them in two ways. First, events are reviewed chronologically to see how the civil engineering profession and its methods have evolved historically. Then, topical areas (e.g., technical methods, management, roles of government and business, education, professional issues) are reviewed to facilitate understanding of the finer points of the civil engineering profession.

## Phases of Civil Engineering

Civil engineering developed in four distinct periods: early civilization to circa 1775, 1775 to circa 1900, 1900 to circa 1975, and 1975 to the present.

### *Early Civilization to Circa 1775*

From early civilization to circa 1775, civil engineering's professional ancestors solved problems of survival, basic systems, and building great structures. Many scientific and mechanical advances were made, the globe was circumnavigated, and basic political organization was initiated. Although methods to solve society's infrastructure and environmental problems were evolving, the term *civil engineer* had not yet been coined.

### 1775 to Circa 1900

The period from 1775 to circa 1900 marked the launch of the Industrial Age, with the steam engine, water power, canals, railroads, and other advances setting the stage for modern society. This period also spans what Joel Tarr, a historian of public works, refers to as the periods of "foundations (1790–1855)" and "constructing the core infrastructure in central cities (1855–1910)" (Tarr 1984).

### 1900 to Circa 1975

This period, about the length of one lifetime, included some earthshaking events: two world wars, the Great Depression, the Cold War, and man's exploration of the moon. Many types of technology were developed, including the telephone, radio, television, computer, automobile, aircraft, submarine, and satellite. The period included the first Earth Day and saw environmentalism become a force. Tarr views in this period the "domination of the automobile and enlargement of the federal role (1910–1955)" and the "rise of the outer city (1955–1982)" (Tarr 1984).

### 1975 to the Present

We are in the Information Age. Technologies such as the Internet have brought on globalization; instant communication; widespread access to knowledge; and new ways to organize and manage organizations, governments, and economies. In some ways this is also an age of limits, with the world's population moving beyond six billion and increasing by another billion every decade or two. Since 1975, global changes, especially climate changes, have become strong factors in civil engineering.

## Origin of the Term *Civil Engineer*

It is important to understand the "genealogy" of civil engineering. As a profession, civil engineering is about 200 years old and shares a common heritage with engineering, science, and management. Early civil engineers were scientists, managers, entrepreneurs, and general engineers; similar to other disciplines, civil engineering began to emerge as a distinct profession during the Industrial Age.

According to John H. Lienhard, winner of the 1998 Engineer–Historian Award of the American Society of Mechanical Engineers, the word *engineering* is probably derived from the Latin *ingeniatorum*. In 1325, a builder of siege towers was called by the Norman word *engynours*, and by 1420 the word in Old English had become *yngynore*. By 1592, the word *enginer* described a designer of phrases (or wordsmith). Later, variations of the word (e.g., *enginer, ingenier,* and *engyner*) were used to refer to builders and makers of things (Lienhard 1998).

During the eighteenth century, engineers began to specialize. The term *civil engineer* was coined in 1768 by British engineer John Smeaton (1724–1792) to describe members of the profession who served the civilian population (as opposed to engineers who served military needs). Also during this time, engineering emerged as one of the "free professions," meaning its practitioners worked apart from organizations such as the church or government. British engineer James Brindley developed methods of independent practice that emerged as consulting engineering (Watson 1988). Today, civil engineering still has many sole practitioners; however, most civil engineers work within organizations.

## Early Civil Engineering

Although the term *civil engineer* has only been in use for 230 years, humans have designed and built structures and systems for thousands of years. A tree trunk or pile of rocks placed across a stream may have been the first bridge. A ditch starting at that stream may have been the first water supply system for a human settlement. The size and complexity of civil works constructed to serve society have grown from such humble beginnings to the comforts of modern life and physical features of our cities and landscapes today.

Antiquity shows that our ancestors performed great feats of engineering. The earliest records show such majestic public and private constructed facilities as the Pyramids of Egypt, the Greek City-State, the temples of biblical times, the aqueducts of the Romans, and the palaces of the ancient Chinese.

Early buildings were built for functional and ceremonial purposes. Early ceremonial structures (e.g., Stonehenge) show solutions to significant construction problems. Stonehenge was built in three phases, from about 2800 to 1500 B.C. Its ring of large stones may have been used for religious ceremonies until about A.D. 43 when the Romans conquered Britain and abolished religious practices (*World Book Millennium 2000* 1999).

The great civilizations of Egypt and Mesopotamia flourished from 4000 to 2000 B.C. Stone, copper, and bronze were used in construction during that period, and many engineering marvels (e.g., the wheel, sails, writing methods) were invented (Oakes et al 1999).

**Figure 2-1**  The Parthenon.

The Pyramids were built from 2700 to 2500 B.C. The Egyptian temples, some of which still exist as ruins today, were of monumental size, as were the Greek temples. The height of the lighthouse marking the entrance to the harbor of Alexandria has been estimated to be 250 ft high, and it stood for many years until it was destroyed in an earthquake.

Civil engineers performed forensic analysis to learn how the Great Pyramid at Giza, Egypt was constructed (Smith 1999). It is roughly two-thirds the size of the Hoover Dam and was originally 481 ft (147 m) high, but the top 30 ft (9 m) have been lost. Civil engineers have been able to estimate how construction problems were solved and were awed by the way ancient Egyptian engineers were able to overcome such technical and management problems as how to produce and move huge stones.

In Greece, the Parthenon (Fig. 2-1) was built from about 447 to 432 B.C. Designed by Greek architects and sculptors, the Parthenon serves as a famous early example of ceremonial building design and construction. It was constructed of marble on a limestone base, on top of a previous structure. Although early stonework was often done without mortar, the Romans discovered natural cement and used mortar in much of their construction, including the Coliseum, which seats about 60,000, many protected from the warm sun by a retractable membrane roof.

The Pantheon was completed in Rome in about A.D. 126. Its name stems from the Greek *pantheion*, meaning "place for all gods." This circular buiding constructed largely of brick and concrete measures about 142 ft (43 m) in diameter. Its dome roof rises about 142 ft (43 m) above the floor at its highest point and was for a time the world's largest known dome (*World Book Millennium 2000* 1999).

The Romans also built impressive roads and aqueducts. The Appian Way was built about 312 B.C. and stretched initially 142 miles from Rome to Capua. By A.D. 200, the Roman road-building era was over. Roman roads were engineered with sand bedding, cement materials, stone base courses, concrete, and crowns for drainage. The entire road cross section could be 2 to 5 ft thick, depending on the foundation underneath. After Rome was conquered, road building languished for centuries.

When Rome was at its peak, engineers built aqueducts and water supply systems. There were, of course, no treatment plants or regulatory agencies. Water was transported long distances to residences and public baths, including the Baths of Caracalla in Rome, which could accommodate about 1,500 persons. Aqueducts were often massive structures with masonry arches. When Rome was sacked by the Barbarians, the water supply system was destroyed, effectively ending the high quality of life in this city for hundreds of years.

Some drainage for cities was used in early times, but the use of sewers to remove waste is comparatively new. A bathroom constructed for King Minos of Greece was purported to have a latrine that was flushed by rainwater (Fair, Geyer, and Okun 1966).

Less known in the West are the ancient public buildings and works of the East, but some are famous, such as the Great Wall of China (Fig. 2-2). Initiated during the Ch'in Dynasty to keep out the Mongolian invaders, it reached a length of 3,080 miles in ancient times and is now about 1,700 miles long.

Impressive engineering feats were also accomplished in the Indus Valley. About 4,500 years ago, civilization flourished in what are now western India and Pakistan. Archaeologists have discovered ancient cities there whose inhabitants had systems of counting, measuring, and weighing. About 1700 B.C. the Indus Valley civilization broke up, and scholars

**Figure 2-2** The Great Wall of China near Beijing.

believe that floods and river changes may have been the cause. Today, civil engineers continue to work on river changes and floods there.

Technology in Europe languished from about A.D. 400 to 900, the period known as the Dark Ages (or early Middle Ages). During this time, civilization sank low and knowledge survived only among religious groups and some palace schools. Public health was abysmal, with many of the problems attributed to contaminated water supplies.

Despite the conditions in Europe, technical advances continued among other civilizations, such as in the Byzantine Empire and among the Arabs. The Byzantines built on Greek and Roman culture, law, and government. They continued Roman traditions of public baths and swimming pools, and chariot races were a form of entertainment. All these activities required public facilities and arenas. Constantinople was a trading center because of its port, necessitating urban and port engineering. They had organized laws, and their Justinian Code became a model legal code. The Emperor Justinian built Hagia Sophia, the large Christian church in today's Istanbul (*World Book Millennium 2000* 1999). During this period, Gothic masterpieces were constructed in Europe without the cement mortar known to the Romans, which was a mystery material to the master builders of the times. These cathedrals illustrate that it is possible to build noteworthy construction with the tools of experience, experimentation, and evolution.

Practitioners of engineering were scientists and technologists who applied their knowledge to practical problem solving. For example, Leonardo da Vinci (1452–1519) created impressive structures and devices. This remarkable individual functioned as an architect, artist, sculptor, and civil and mechanical engineer.

Although there have been numerous feats of civil engineering in Europe, Africa, and Asia, the impressive feats in the Western Hemisphere should not be ignored. For example, the Incas created Machu Picchu, an archaeologic site in Peru that served as a royal estate. This site contains ruins of buildings constructed of granite in the 1400s, with steep thatch roofs for drainage in the wet climate. The Denver civil engineering firm Wright Water Engineers has obtained an archaeologic permit to investigate the site and has initiated a project to study the water systems at Machu Picchu (Wright, Zegarra, and Lorah 1999).

By 1500 much engineering activity was in progress. The Renaissance (the period in Europe between the Middle Ages and the Enlightenment) renewed interest in Greek and Roman cultures. In Italy, wealthy Florentine families commissioned large, beautiful palaces and religious and civic buildings. It was a time of artistic advances, including beautiful buildings designed by Renaissance architects. This period also featured trips by explorers such as Christopher Columbus, John Cabot, Giovanni da Verrazzano, and Amerigo Vespucci. America was opened for exploration by Columbus' famous trip, and mysteries of the New World started to unfold.

Next came the Enlightenment (or Age of Reason), a period that stressed the use of reason to learn truth. Lasting from the 1600s until the late 1700s, this period is also known as the Age of Rationalism because engineers of the time used reason and rational thinking. Leaders of the period relied on the scientific method, with emphasis on experiments and empiric observation. Physical sciences were important, including the laws of gravity and motion formulated in England by Newton.

During this time, the French contributed a great deal to the development of civil engineering (Lienhard 1998). By 1650, military applications of artillery and fortifications required officers to know mathematics and mechanics. This was the beginning of military engineering. King Louis XV authorized a School of Bridges and Highways in 1775 and, after the French Revolution, Napoleon replaced it with the École Polytechnique. Famed engineer Jean Baptiste Joseph Fourier, born in France in 1768, taught there in 1795. In 1798, Fourier traveled to Egypt with Napoleon to build roads. Napoleon's engineers built several bridges with varying success. They built a 106-ft arch over the Seine, naming it after the Battle of Austerlitz. France trained a generation in applied analysis. The names of 72 important scholars who founded the science and mathematics of structures are written in the Eiffel Tower.

## The Growth Period of Civil Engineering

Up to this point in history, many remarkable technologic advances had been achieved, many of which were the result of techniques that became known later as civil engineering. The biggest advances, however, were yet to come.

### *1775 to 1900*

Much of civil engineering development coincides with the emergence and growth of the United States as a nation. Between about 1775 and 1900, dramatic political and technologic developments shaped civil engineering. Important developments were made in both the United States and throughout the world. By the 1700s, America was a colony of England, France, and Spain. New settlers required energy and technologic systems. By the eighteenth century, engineers were beginning to specialize, and the term *civil engineer* was coined in 1768. At about the same time, British engineer James Brindley developed methods of independent practice that emerged as consulting engineering (Watson 1988). At this time, the practice of engineering was performed mainly by entrepreneurs rather than by firms.

Engineering societies and organizations were emerging. In 1771, exploratory meetings to charter the Institution of Civil Engineers were held in Britain, and this group was chartered in 1818. Meanwhile, the first US engineering school was organized at West Point in 1802 with assistance from French engineers. Civil engineering societies were organized in Holland (1843), Belgium (1847), Germany (1847), and France (1848); and the ASCE was chartered in 1852.

At the beginning of the nineteenth century, the American Revolution was over but Napoleonic wars were still raging in Europe. The United States was becoming a nation, significantly increasing its area with the Louisiana Purchase. Lewis and Clark explored the land to the Pacific Ocean. There were few urban centers, and even such metropolitan areas as New York had small populations (in 1800, New York City had a population of only 60,000). The western and southern parts of the United States were largely unsettled. Infrastructure was limited, but such technologies as road building were known in Europe, having been rediscovered during the Enlightenment.

At this time, few people identified themselves as civil engineers. The census of 1850 counted only 512 civil engineers in the United States. People had to be independent and practice different skills. George Washington functioned early in his career as a civil engineer and surveyor. He was the first of several presidents with an engineering background. Others included Herbert Hoover and Jimmy Carter.

By 1820, the Industrial Revolution was beginning. The use of steam engines increased for transportation and other purposes. Canal transportation increased, and the railroad was invented and developed. The Erie Canal was sometimes known as the first American "university of civil engineering" because engineers of the early 1800s learned their profession as assistants and apprentices on that project. Although England had impressive canals, canal building in the United States developed more slowly due to lack of capital for development. Imagine the challenge: to build the longest canal in the world—364 miles—through largely unsettled wilderness. The New York State legislature, under Governor Dewitt Clinton, authorized the project despite the lack of federal financial assistance. Begun in 1817, the canal was completed in 1825 and was so successful that the $8 million construction cost was recovered from fees in just 7 years. Freight charges from New York City to Buffalo dropped from $100 to $10 per ton, and the time required to move a load decreased from 26 to 6 days (Armstrong 1976).

Public health was poor during this period, and recognition of the role of civil engineering was low and remains so today. For example, Florence Nightingale gained fame as a nurse during the Crimean War in about 1850, but the engineer who was knighted by Queen Victoria for the role played by the British Army's water supply enjoys less fame.

As a growing nation, the United States needed water and power, especially to fuel growth and industry as cities grew on the fall lines, or the zones where hill country changed to plains. During this period, public works such as canals, watermills, harbors, bridges, and dams were being designed and built by civil engineers (Wisely 1974, Armstrong 1976). Towns in Massachusetts and other New England states needed water power to run textile mills. Later, abundant labor was available in the South, and industry moved to states such as North and South Carolina, which also required water power. Civil engineering was becoming a discipline, with its own body of knowledge and professional standards of practice. Surveying needs grew as new lands were settled, and surveying became a branch of civil engineering.

The 1840s saw the Mexican War, expansion into the West, and the California Gold Rush, all of which created a need for water diversion structures, flumes, dredges, and hydraulic machinery. As the West opened to mining, settlement, and agriculture, the Mississippi River works were under way and the first bridges were planned. Westward expansion was fueled by federal land grants to railroads. In 1869, the Union Pacific and Central Pacific rail companies completed the first transcontinental route (with the famous golden spike ceremony) at Promontory Point, Utah. In other parts of the world, civil engineering activities included projects such as the Suez Canal (built in the 1860s).

After the Civil War, the nation recovered and grew. Cities gradually developed public works organizations but were reluctant to form operating departments or to incur debt for services. This led to haphazard public–private development of infrastructure. Road maintenance was often a shared activity among farmers who lived along the roads. In the last decades of the nineteenth century the light bulb, telephone, automobile, and assembly line were invented. The Industrial Revolution created new demands for infrastructure, and the number of civil engineers in the United States increased to about 20,000 by 1900. The automobile was about to change the form of cities and roads forever.

## *1900 to 1975*

The increasing availability of cars, especially between 1910 and 1930, required streets and road systems beyond previous expectations. World War I placed great stresses on existing infrastructure systems and necessitated new roads in the United States to transport war material. In 1927, Charles Lindbergh made the first trans-Atlantic flight. Trans-Atlantic cables and radio began to connect the world as never before.

The 1930s saw the Great Depression and dust bowl conditions, thereby creating a greater need for civil engineers. The National Industrial Recovery Act of 1933 called for a comprehensive program of public works (Holmes 1972). Senator George Norris of Nebraska championed the development of the Tennessee Valley Authority (TVA) and federal projects in Nebraska, which became a pure public power state.

The need for all types of engineers increased during World War II. Corps of Engineers' General Leslie Groves was responsible for logistics of the Manhattan Project, which created the first atomic bomb. The Flood Control Act of 1944 authorized the Pick-Sloan plan for coordinated and comprehensive development of the Missouri River Basin. It authorized projects for irrigation, power, and flood control and provided for use of the projects for recreation as well.

The rising standard of living after World War II mandated new cities and infrastructures. For example, Levittown, Pennsylvania was built to provide housing for the masses, introducing the subdivision to American life. Single-family housing increased the need for water, energy, and transportation. Management organizations were created to meet the needs of infrastructure operations. With suburbs there was a big increase in the number of special-purpose districts for infrastructure. Central cities grew to meet them, and the growth in the number of special districts was slowed by 1970.

Water became a big issue after World War II, and in the 1950s a Senate Select Committee on Water Resources was appointed. This led to the Water Resources Planning Act, first passed in 1962 (Holmes 1979). The 1960s also saw the growth of the Space Program; a rise in environmentalism; and the Great Society, which included a New Cities Program.

Some saw government programs of this era as the "last gasp of the New Deal." For example, 1965 to 1980 was an active period in water resources planning, but activity diminished greatly after the 1980 presidential election. The political turmoil of the Vietnam and Watergate eras dominated government policy during that period.

During the 1970s, environmental legislation increased dramatically. Laws such as the National Environmental Policy Act, the Clean Water Act, the Clean Air Act, and the Endangered Species Act added new dimensions to civil engineering. Another important aspect of the 1970s was the energy crisis and OPEC oil embargo.

## *1975 to 2000*

In 1981, Pat Choate and Susan Walter (1981) launched the "infrastructure crisis" with the publication of *"America in Ruins: Beyond the Public Works Pork Barrel"* (Choate and Walter 1981). The 1980s also saw breakup of the telephone monopoly and large-scale privatization in Britain, including parts of the water, energy, and transport sectors. The influence of government waned in favor of the private-sector approach to problems. Internationally, the 1980s ushered in such global concepts as sustainable development and focused attention on problems of poverty and water and sanitation in developing countries.

By the 1990s, the Berlin Wall had come down and Europe was integrating its political and economic systems. The United States was experiencing an unprecedented economic boom. The rise of telecommunications and computing seemed to drive the economy and growth-induced demand for infrastructure systems.

The Information Age began around 1975 with the invention of the microcomputer and silicon chip. This era features new challenges to civil engineers that relate in part to the Internet, e-commerce, and new technologies and in part to the aspirations of billions of people for the "good life" they see on television. With anti-tax fever and mistrust of government added to the mix, the emphasis is on tax cutting, regulatory reform, privatization, and nongovernmental initiatives. Both major political parties in the United States embrace this agenda.

## Areas of Civil Engineering Practice

As technology and civilization unfolded, areas of civil engineering developed, gradually forming the diverse specialties that exist today under the umbrella of the profession.

### *Construction Technology*

Construction technology blends the use of machines, materials, energy, and management techniques. As knowledge increased through the Industrial Revolution, humans became better and more efficient at building things. Humans had already proven they could construct majestic structures; it was simply a matter of building larger and more complex structures.

The invention of the steam engine in the nineteenth century provided the capability of moving large amounts of earth with steam shovels. This invention gave way to mammoth gasoline- and diesel-powered construction machinery. Wartime exigencies, along with such megaprojects as the Suez and Panama Canals, led to the development of new technologies to solve such problems as scaffolding, large-scale site dewatering, river dredging, and demolition.

Today, the United States construction industry is a $750 billion per year enterprise with some two million firms that is able to tap expertise from all sectors of the economy to solve extremely complex technologic and management problems. Civil engineers are active in the industry but share its work with other key disciplines, such as construction managers and technical specialists in different types of construction.

### *Buildings and Structures*

Construction technologies, combined with new and rediscovered materials such as wrought iron, cast iron, and natural cements, have led to improved buildings. The birth of the steel industry in the nineteenth century enabled engineers and architects to plan new types of buildings and frames. Steel frames, along with the elevator and a growing density of cities, led to development of the skyscraper.

Some consider elevator buildings, which appeared in New York during the 1870s, to be the first skyscrapers. An example is the seven-story Equitable Building, which was built in 1870 and had two steam-powered elevators. The Tribune Building and the Western Union Building are better known for their heights of 10 stories. The Home Insurance Building in Chicago, built in 1884 and 1885, was the first to have a complete frame as a supporting structure. The Woolworth Building, completed in 1913, brought key skyscraper technologies together for the first time. The building was supported on concrete piers that extended to bedrock and were constructed using caissons. Every five stories the building has an additional smaller story for water tanks that fed lower floors. The Woolworth Building has a steel frame that is carried up to 792 ft, which was very tall for the time. The building's system of wind bracing is highly developed, and it has high-speed elevators.

The five tallest buildings in the world as of 2000 are Petronas Towers, Kuala Lumpur (1483 ft), constructed in 1997; Sears Tower, Chicago (1450 ft), constructed in 1974; Jin Mao Building, Shanghai (1379 ft), constructed in 1998; World Trade Center, New York (1368 ft), constructed in 1972; and Plaza Rakyat, Kuala Lumpur (1254 ft), constructed in 1998. The Empire State Building, once the world's tallest at 1250 ft and completed in 1931, is still sixth in the world (History Channel 1999).

## *Transportation*

During the Enlightenment new roads were built in France and England, and the art of road-building was restored. Europe also developed laws governing road construction. Early Saxon law, for example, imposed three duties on lands: to repair roads and bridges, maintain castles and garrisons, and repel invasions (Oglesby 1975). Development of roads and bridges in the United States began with the arrival of early settlers from Europe. These settlers brought with them their newly rediscovered arts of road and bridge building, although initially roads consisted of primitive trails and farm roads that turned to mud in wet weather. There were a few exceptions. For example, the Whiskey Rebellion in 1794 led to construction of the Philadelphia-Lancaster Turnpike, which was a 62-mile toll road surfaced with hand-crushed stone and gravel.

In cities, cobblestone was used for street paving; but its associated noise, fragility, and difficulty in cleaning rendered it unsatisfactory and led to the search for better materials. It was not until the beginning of the automobile era that modern asphalt and concrete pavements began to be used. Roads improved somewhat in the nineteenth century, but it took the automobile to create the demand that led to today's network of paved highways. Around 1910, Henry Ford's Model T and "Doc" Kettering's electric self-starter made cars affordable and friendlier. In the first two decades of the century millions of automobiles came into use, creating enormous demands for roads. Then, in the twentieth century, the automobile and road transportation became the forces that shaped modern cities.

The automobile was transformed from a curiosity to a mainstay of American life during the first three decades of the twentieth century. Cars created personal mobility, allowing cities to spread into suburbs from dense inner-city residential areas (as are still seen in many European and some older US cities), and opened a need for new transportation networks (i.e., highways rather than railroads). This beginning of urban sprawl is a subject of debate today, as some see good and some see evil in dispersed rather than compact living. The debate is played out under the rubric of smart growth.

When the US highway system was introduced, the first highways were named after people (e.g., Lincoln Highway); later, numbers were used (e.g., Highways 30, 66, 101). During the Eisenhower administration, with a view toward military transportation as well as other public needs, interstate highways were initiated. This system of highways now covers more than 45,000 miles, and its official name is the "Dwight D. Eisenhower System of Interstate and Defense Highways." Figure 2-3 shows the scale at which interstate highways have affected urban areas in the United States.

Transportation now requires mega-projects, such as Boston's Big Dig, which aims at massive upgrades in traffic corridors and systems, including tunnels and bridges built within the confines of the city. Bridges got longer as projects addressed water features and canyons not previously bridged, and airports got bigger and harder to locate as air traffic increased and open areas near cities became more limited. Recent mega-projects included the Japanese Honshu-Shikoku system of bridges (see Chapter 3), reconstruction of the San Francisco–Oakland Bay Bridge, the Denver International Airport, the Hong Kong Airport, and other mass transportation projects.

While the nation was waiting for rail and auto transportation to develop, it turned to canals to move goods. Construction of the Erie Canal was the signal achievement in early nineteenth-century America. Later, large-scale canal building focused on the Suez Canal in the 1860s and then the Panama Canal in the 1880s to early 1900s. Many construction, geotechnical, and hydraulic problems were solved during the course of these construction projects.

Railroad building was another impressive American engineering achievement in the nineteenth century. Again, knowledge had originated in England with experiments from 1800 to 1820. A rail line opened in 1825 between Stockton and Darlington, England, and American devel-

**Figure 2-3**   Interstate highway system through Denver. (Courtesy Colorado Department of Transportation; photo by Gregg Gargan)

opers began proposing lines for construction. In 1820, Oliver Evans proposed a line between New York City and Philadelphia, but it was not until 1827 that the first commercial American line, the Baltimore and Ohio (B&O) Railroad Company, was chartered in Maryland. British engines were unsuited for rough American conditions, and the B&O finally developed a practical locomotive in 1831. The unloaded speed of 30 mph was reached in 1830 by the South Carolina Canal and Railroad Company.

Railroad expansion faced many obstacles. Construction of the Hoosac Tunnel in the Berkshire Mountains in New England is an example. Construction of this tunnel lasted from 1848 to 1875, cost $20 million (10 times the original estimate), and claimed the lives of 195 men. At the time, it was the nation's longest tunnel and one of the engineering marvels of the nineteenth century (*World Book Millennium 2000* 1999).

Early railroad bridges represented a great leap forward from other bridges that served horse-drawn vehicles. Variations of metal trusses were developed during that time, with many of the designs and systems being patented. An example is Eads' bridge across the Mississippi River at St. Louis. Completed in 1874, the bridge was the first major structure in the world to use steel in its primary structural elements (Armstrong 1976). Bridge building evolved during the nineteenth and twentieth centuries into an impressive art and science, and many new designs were developed.

Aviation developed rapidly after 1900. The Wright brothers' plane evolved into passenger planes, which required airports and other infrastructure systems to serve a growing industry. Today, the aviation industry combines many technologies and skills; and the number of passengers flying annually is increasing rapidly.

## *Water, Wastewater, and Environmental Engineering*

Water and wastewater engineering have evolved to solve many problems of convenience, economics, and health. Today, engineers working in this field are called environmental engineers, and their practice has evolved to embrace other problems of public health and ecology.

The earliest recorded water supply system in the United States dates from 1754 when the Moravian settlers of Bethlehem, Pennsylvania built a system of spring water forced by a pump through bored logs. The first sand filters were used in London in 1829. Hamburg got a public water supply in 1848. Helsinki's supply system dates to 1876, and the first wastewater treatment in Helsinki was put in place in 1910 (Katko 1997).

In Europe and in inner cities of America, public health was abysmal during the nineteenth century. Records of sanitary facility management have been published, such as Charles Dickens' writings about 1850s London in which people lived together without their basic sanitation needs met. Doctors, lawyers, engineers, and other professionals prodded the social consciences of the people and their governments, eventually leading to advances in public health.

Wastewater systems began with in-house water supply systems, which overloaded cesspool and privy vault arrangements. Combined sewers emerged as a solution to carry away domestic waste products, but these systems created their own pollution problems. Storm drains provided convenient receptacles for waterborne wastes, and with the invention of the water closet the current wastewater system was in place. The discovery of the water seal for building drains finally made in-house plumbing socially acceptable (before the use of seals the odor was unacceptable).

The idea of expending public funds for water services for all people is comparatively new. People living on farms could drink from wells and dispose of wastes freely. Although they had their own health problems, water and wastewater disposal were not central issues. Slums, with concentrations of people living in urban poverty, brought new problems to urban areas. Water was drawn from polluted rivers and excrement returned to them via drainage into the rivers. Although there were rules against placing waste in storm drains, it happened anyway and serious public health problems resulted.

Philadelphia initiated its water supply system in 1798 after a yellow fever epidemic. The system combined public and private pumping facilities, with pumps driven by horses. In these early days of water supply in America every city had a unique story (Armstrong 1976). By midcentury, other large cities such as New York and Boston followed Philadelphia's lead. Systems were both public and private until after the start of the twentieth century, when the balance shifted to public ownership. Some private water companies remain, and today privatization is popular (see Chapter 10).

The period around 1850 in the United States included a renaissance in thinking about public health and laid the foundation for advances against waterborne pollution later in the century. Well-known engineers of the time began to devise plans to improve health by such measures as bringing in clean water, building drain lines, and carrying out research on sanitation engineering (Fair, Geyer, and Okun 1966).

The nineteenth century witnessed advances in halting waterborne disease. Factors included protection of water supplies, improvements in wastewater disposal, beginning of regulation, and an awareness of sanitation and public health. Sewers were developed to remove wastewater from central cities, and pressure mounted to use storm sewers for sanitary wastes. This led to the combined sewers seen in large cities in the United States and Europe today. Finally, connections between water supplies, water courses, and wastewater were recognized. Communities were served by springs and wells that were easily contaminated. New York City went a great distance for water with the construction of their Croton Aqueduct, which opened in 1842 (Armstrong 1976).

As the twentieth century dawned, the United States faced such problems as cholera and typhoid. These diseases were eventually eradicated through environmental engineering and public health measures, leading to a greater appreciation of the role of environmental engineers in public health. During the early twentieth century, water supply and wastewater regulatory programs emerged. The United States Public Health Service was created, with parts evolving into today's Environmental Protection Agency (EPA). After World War II, interest in water and wastewater regulation increased, and legislation was passed to create the current system. For the most part, policies relating to stormwater have evolved more from concern about water pollution than from local drainage and flood control. Solid wastes and air quality were added to water supply, wastewater, and stormwater to form the specialty of environmental engineering.

At about the turn of the century, the nation entered a conservation era, with the creation of national forests and parks. Environmentalists such as John Muir received attention and initiated a movement that peaked with Earth Day in 1970. This added an ecologic thread to environmental engineering, and today's engineers dealing with fisheries, environmental impact, and wildlife may also call themselves environmental engineers. Environmental management has also evolved as a field and is distinctly interdisciplinary, including a number of science and management areas.

Figure 2-4 shows an environmental engineering feature of the 1990s. South Florida's Kissimmee River has been channelized, and its bends and natural features are being restored.

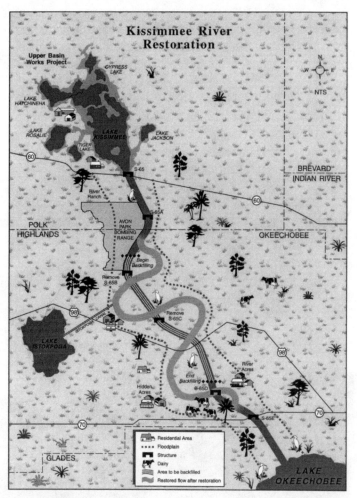

**Figure 2-4**  Kissimmee River restoration, Kissimmee, Florida. (Courtesy South Florida Water Management District)

## Water Resources Engineering

Civil engineers develop techniques to handle water for energy, flood control, navigation, and recreation. This requires a multipurpose approach, and the field of water resources engineering emerged separately from those of environmental engineering. One of the early water resources bills, the Flood Control Act of 1917, called for a "comprehensive study of the watershed," including the study of power possibilities. However, the Corps of Engineers, the nation's main flood control agency, was resistant to the concept of multipurpose planning (Holmes 1972). Later, multipurpose and comprehensive planning became the norm.

Major twentieth-century water projects included high dams such as the Hoover Dam, TVA projects (with great social impact in affected areas), and Bonneville Power's Grand Coulee Dam. The Corps of Engineers and the Bureau of Reclamation are generally considered the nation's dam builders, along with power companies and especially the TVA. Dam building declined after the 1960s, and dams are now generally considered a mixed blessing. Dams can provide tremendous economic benefits, but they also have a significant environmental impact.

Figure 2-5 shows the Guri Dam in Venezuela, which supplies large quantities of power. This dam was planned and designed by Harza Engineering Company, with leadership from Victor A. Koelzer, who joined Colorado State University later and was elected an Honorary Member of ASCE in the early 1990s.

**Figure 2-5**  The Guri Dam in Venezuela.

*Disaster and Emergency Management*

As civil engineering has evolved, it experienced failures and emergencies. This led to interest in such fields as risk management, forensic engineering, and safety engineering. The nineteenth and twentieth centuries saw many horrific disasters whose consequences were worse because engineering methods and policies had not been sufficiently evolved. For example, the Chicago fire of 1871 burned for more than 24 hours, wiped out the downtown, killed at least 300 people, and left 90,000 homeless. The 1900 Galveston hurricane killed about 6,000 people and was the worst natural disaster in US history. The 1906 San Francisco earthquake destroyed about 28,000 buildings and killed more than 3,000 people (*World Book Millennium 2000* 1999).

The field of disaster management has evolved to manage risks resulting from natural and man-made disasters. Vulnerability was recognized as a key issue because losses are much greater when populations are vulnerable. For example, a 1970 cyclone in Bangladesh killed about 266,000 people. Disaster management has also evolved to handle new threats. For example, the 1986 Chernobyl nuclear plant disaster near Kiev in the Soviet Union showed how vulnerable societies are to nuclear threats. Today, the list of disasters to protect against has grown much longer to include various acts of terrorism as well as greater natural disasters.

## Development of Formulas and Methods

Part of the heritage of civil engineering lies in the development of formulas and methods used for studies and designs in our separate technical lines (e.g., structures, geotechnical, hydraulics, hydrology, sanitary engineering). For example, in the field of structures, the methods developed by Hardy Cross to analyze frames were later extended to the analysis of pipe networks. In hydraulics, major empiric advances were made in the nineteenth century, including formulas for pipe friction, open channel flow, and movement of waves in channels. In hydrology, the Rational Method was developed, setting the stage for the extensive data management programs that are currently in use. In sanitary engineering, the understanding of biology and chemical nature of water pollution emerged in the past century along with the fields of public health and medicine. The technical methods of each specialty within civil engineering have an interesting history.

## Evolution of the Field of Management

While some engineers were creating technical methods and formulas, others were contributing to the field of scientific management. Perhaps the most famous engineer–manager is Frederick W. Taylor, who practiced around the beginning of the twentieth century, developing such quantification methods as time and motion studies. Taylor sought to improve processes so they would be scientific (that is, not subject to the whims of human nature) and to make management itself a science. Although the field has evolved to include many other sectors, Taylor's early contributions form the origins of quantitative management and show how engineers make contributions to the field of management.

As civil engineers increasingly accepted management roles in infrastructure systems, they found increases in complexity, in the political content of problem solving, and in the way civil engineering was practiced. Working with government processes required civil engineers to learn about public administration. Infrastructure and environmental problems have always involved politics because they affect so many people and are in the public arena. How public decisions are made depends on how democracy and government work. Today, with ballot initiatives and Internet access, there is a trend in America toward mass democracy rather than representative democracy. This affects the way that civil engineers get involved, for example, in capital improvement projects.

The trend is evident in citizen initiatives. Formerly, elected representatives would make decisions; now, the public becomes involved more directly. The past featured emphasis on a few powerful decision makers; today the public must be sold on the alternatives. The media is also deeply involved.

Mega-projects, such as the Big Dig in Boston, are being built, but there is often emphasis on less capital-intensive approaches, such as demand management in water and transportation systems. Regardless, tomorrow's great population increases will continue to require large projects. Environmental and social impact analysis has increased the complexity of planning and the number of experts who get involved. Today's political climate requires a search for workable processes. This is not the result of a particular project, place, or set of players; rather, it is directly connected to the evolution of our political institutions.

Water planning is an example. In earlier years, water resources planning meant planning for economic goals such as hydropower or irrigation. The capital investments caught the attention of the public and had political support. During the twentieth century, the goal became comprehensive planning and multiple purposes and several players were involved. As society has grown more complex, so has the scope of comprehensive planning, which now means to consider practically every aspect of society. This occurred because as the nation developed it became clear that water involves many industries, geographic areas, and public interests, and the possibilities of multipurpose development have become apparent.

## Local Government: South Florida Water Management District

Taking care of streets and sanitation has led many civil engineers to work in local government, including towns, cities, counties, and special purpose districts. Each of these divisions has its own history. The story of the South Florida Water Management District (SFWMD) illustrates this. There are many units of local government (over 80,000 in the United States, including school districts), so many stories about local government can be told.

The extreme subtropical weather of South Florida, including hurricanes, floods, and droughts, led Congress to create the Central and Southern Florida (C&SF) Flood Control Project in 1948, which led to the creation by the Florida Legislature of the C&SF Flood Control District. This became today's SFWMD, which operates and maintains some 1,800 miles of canals and levees, 25 major pumping stations, and about 200 larger and 2,000 smaller water control structures. The District serves 16 counties with a total population of about six million people, covering 17,930 square miles of agricultural lands, water conservation areas, and urban areas. Its budget is funded by property taxes and other sources such as federal and state revenues as well as grants. In fiscal year 2000, the SFWMD budget was $469.1 million.

The District has over 900 employees, including field engineers and inspectors, to oversee surface and groundwater systems. Land management professionals help manage pristine natural areas and recreational sites. They also employ planners, engineers, and environmental scientists for regional water supply planning; surface water improvement management (SWIM) of Everglades restoration research, monitoring, or construction projects (SFWMD 1999). The history of the SFWMD parallels important engineering and environmental developments of the United States, and it is an example of the many local government opportunities for civil engineers today.

## State Government: California Department of Transportation

The emergence of automobile transportation gave rise to one of the largest employers of civil engineers: the state highway department. State governments also employ civil engineers in environmental and other work, but the majority of government civil engineers work in transportation. The growth of civil engineering in state governments is evident from the history of the California Department of Transportation (CALTRANS).

By the 1850s, miners and merchants had woven a rough network of supply roads through California. California was one of the first states to have a Bureau of Highways Commission, which met for the first time in 1895 in Sacramento. Only 1.5 years later, the Commission recommended a 14,000-mile highway network, which was intended to link agricultural areas and population centers. Following World War I, gasoline taxes began to be used to fund highway construction, operation, and maintenance; and the American Association of State Highway Officials began to establish national design standards.

The farm-to-market road system, started in the 1930s under the administration of the US Bureau of Public Roads, led to an expanded system of paved rural roads. The Pacific Coast Highway was completed in California in the 1930s. The 1940s issued in the freeway era and preceded development of the interstate highway system. Changes in the 1960s and 1970s led to the new name, California Department of Transportation, and the trend caught on around the nation. CALTRANS emphasized the efficient use of highways and their integration with other modes, such as rail and transit, carpooling, ramp metering, telecommuting, flexible work hours, and intelligent transportation systems (CALTRANS 1999).

## Federal Government

Civil engineers working for the federal government include those in the military and on the civil side of the Corps of Engineers, as well as the Bureau of Reclamation, Department of Transportation, EPA, Federal Emergency Management Agency, Department of Energy, Department of Agriculture, and others.

A view of federal involvement in civil engineering can be seen from the web site of the House Transportation and Infrastructure Committee (the "Building Committee" of the US Congress). This committee, whose roots reach to the beginning of the Republic, traces interesting aspects of civil engineering history:

> In the early life of the Nation, public works were linked directly and almost exclusively to the needs of commerce—roads, inland waterways, coastal harbors, and navigational aids that served to foster trade between the States and with the merchant nations of the Old World. In fact, the first recorded public works project in the United States, authorized by the First Congress in 1789, was a lighthouse on Cape Henry to guide shipping in and out of Virginia's Hampton Roads. At the outset, there was serious disagreement over the constitutionality of federal participation in such internal improvements (as public works originally were termed), and it was

not until 1824 that Congress passed a General Survey Bill specifically permitting federal assistance to state projects. With the enactment of this law, the Public Works Program of the United States was truly under way and a stream of federal funds began flowing into national improvements. The early concentration on waterways and roads soon broadened into new and more sophisticated forms of public works necessitated by the nation's westward expansion and its growing interest in the markets of Europe and the Far East. Federal land grants, eventually amounting to almost 10% of the land area of the United States, were given to the railroads to help establish new routes and to link the furthermost sections of the country. Concurrently, a vast overhaul of our eastern rivers and harbors was undertaken to facilitate the growth of our increasingly important maritime commerce. By the dawn of the twentieth century, the 13 original States had grown into a continental Union with problems and potentials undreamed of by the founding fathers. The arid lands of the American West lay waiting for settlement, but they would remain forever barren and useless without proper irrigation on a massive scale; the mighty Mississippi and its tributaries offered limitless promise for agriculture, commerce, and industry; but the promise was thwarted by devastating floods that returned again and again to blight the land. With the new century came the realization that the Western Desert must be made fertile and the rivers harnessed; and so a new era of public works began. In 1902, the Federal Reclamation Act was passed, placing revenues from the sale of public lands in a fund to finance federal irrigation projects. The prosperous ranches and farmlands of western America today bear witness to the success of this venture, which has become a multi-billion-dollar continuing program under the direction of the Interior Department's Bureau of Reclamation. In 1917, Congress turned its attention to the recurrent floods that for centuries had ravaged the great river valleys of America's heartland. The result was the first Mississippi Flood Control Program, which was greatly expanded after the disastrous flood that inundated the Mississippi Valley in 1927 and enlarged again in 1936 into a comprehensive program embracing all of the nation's major streams. Under the direction of the Public Works Committee and its predecessors in the Congress, the United States Army Corps of Engineers built dams and flood control projects throughout the country. The Committee had foreseen the need for similar dams and flood control projects along the Susquehanna and Potomac Valleys and had authorized their construction. Had these systems been built; the tragedy and loss of life and property wrought by tropical storm Agnes in June 1972 could have been in large measure averted. In the 1930s, public works were brought to bear against yet another form of disaster—the Great Depression—which almost brought America to a standstill. In order to provide jobs for millions of desperate men and women, the Work Progress Administration (WPA) and the Public Works Administration (PWA) were created, and together they helped bring the nation safely through its crisis, leaving behind them such lasting and useful monuments as schoolhouses, health centers, roads, and the TVA's stupendous complex of electric power, flood control, conservation, and economic development projects (United States House of Representatives, Committee on Transportation and Infrastructure 1999).

Federal government contributions to civil engineering must acknowledge the many pioneers in the military or in federal agencies who developed engineering methods and systems. The Corps of Engineers, for example, has been active in developing projects ranging from the Panama Canal to the Manhattan Project, which led to the first atomic bomb. The Department of Agriculture developed systems for irrigation engineering, including the Parshall Flume, a unique measuring device for water, developed at Colorado State University by Ralph Parshall. The Bureau of Reclamation pioneered water reclamation technologies in the West, and the EPA has sponsored research to develop improved processes for environmental management.

## Early Engineering Education

The first engineering college in the United States was the US Military Academy at West Point. During the Revolutionary War, both sides realized the strategic importance of the west bank of the Hudson River, and General George Washington selected Thaddeus Kosciuszko to design the fortifications for West Point. These included forts, batteries, and redoubts as well as a 150-ton iron chain across the Hudson to control river traffic.

Desiring to eliminate America's reliance on foreign engineers and artillerists, such national leaders as George Washington sought the creation of an institution devoted to the arts and sciences of warfare. President Jefferson established the United States Military Academy (USMA) in 1802 after ensuring that those attending would be representative of a democratic society. Colonel Sylvanus Thayer, "father of the military academy," served as superintendent from 1817 to 1833. He upgraded academic standards, instilled military discipline, and emphasized honorable conduct. Thayer made civil engineering the foundation of the curriculum. For the first half-century, USMA graduates were largely responsible for the construction of the bulk of the nation's initial railway lines, bridges, harbors, and roads (USMA 1999).

The development of other technical schools in the post–Civil War period allowed West Point to broaden its curriculum beyond a strict civil engineering focus. The USMA came to be viewed as the first step in a continuing Army education. Initiation of the program at West Point was followed by programs at other schools, such as Rensselaer Polytechnic Institute (RPI) in New York State and MIT. The Land Grant Act in 1862 led to many of the agricultural and mechanical (A&M) applied universities, which were often assigned the engineering education function. Great growth in the number of engineering programs resulted from the Land Grant Act. The other great period of growth in engineering education came after World War II, when the nation's science base was expanding.

## Organizing for Civil Engineering and Management

The nineteenth and twentieth centuries saw advances in social organization for engineering. One innovation was city management. In the early days of US cities, there was no professional cadre of engineer–managers and decisions were made by citizen governing boards. As the profession grew in the United States, civil engineers began to take on the role of administrative and technical management of infrastructure systems.

Public works grew more complex and required more sophisticated management. Civil engineers, represented by ASCE, had developed tools for public works management by 1900. The American Society of Municipal Improvements (ASMI) was formed in 1894, with the first meeting held in Buffalo. A split in 1897 with the League of American Municipalities led the ASMI to focus on technical matters and then to merge with the Association for Standardizing Paving Specifications in 1913. The new organization was named the American Society of Municipal Engineers (ASME) in 1930. Another organization, the International Association of Street and Sanitation Officials, operated in parallel with ASME. In 1925 it became the International Association of Public Works Officials (IAPWO). In 1935 ASME and IAPWO formed a joint secretariat and created the American Public Works Association, which operates today (Armstrong 1976).

By the mid-1950s, environmental engineers decided that certification would be beneficial for areas such as water supply engineering, air pollution control engineering, and wastewater treatment. They created a Joint Committee of the American Public Health Association, American Society for Engineering Education, American Society of Civil Engineers, American Water Works Association, and Water Environment Federation to found the American Academy of Environmental Engineers. Later, other sponsors joined, including the American Institute of Chemical Engineers, American Public Works Association, National Society of Professional Engineers, Association of Environmental Engineering Professors, American Society of Mechanical Engineers, and Solid Waste Association of North America. These diverse groups illustrate the cross-disciplinary nature of environmental engineering (American Academy of Environmental Engineers 1999).

With the increased complexities of water resources, ASCE created the Water Resources Planning and Management Division in 1973. The division celebrated its twenty-fifth anniversary at its convention in 1998. In 1999, ASCE created a new Environment and Water Resources Institute, which merged these water divisions with the environmental division, creating an integrated entity.

## Engineering Societies and Professional Registration

The ASCE has been a focus for mutual work of consulting engineers since its founding in 1852. Most of the original 12 founders and the early ASCE members, such as John A. Roebling, were private practitioners.

Associations for the business practices of consulting engineering came later. Today's American Consulting Engineering Council (ACEC) arose from a merger of the American Institute of Consulting Engineers (AICE) and the Consulting Engineer's Council of the United States (CEC). AICE was formed in 1910 from leading individual practitioners living mainly in New York City. CEC was founded in 1959 mainly to serve promotional and business needs, as was done by The American Institute of Architects (AIA) (Wisely 1974). In Britain, an Association of Consulting Engineers was established in 1913 (Watson, 1988). Its main focus was on promoting noncompetition and selection based on qualifications, a concern that still exists among consultants despite the governmental ruling that competition must exist. Also in 1913, the International Federation of Consulting Engineers (FIDIC, French acronym) was formed in Europe. In addition to ASCE and ACEC, other engineering organizations such as the National Society for Professional Engineers (NSPE) provide mechanisms to safeguard the practice of consulting engineering.

Early thought was that membership in ASCE was a sufficient credential for civil engineers. During the late nineteenth century there were a number of engineering failures, however, and the first proposal for professional registration was made in 1897 when Wyoming officials learned that people signing maps were lawyers, notaries, and other unqualified individuals. In 1907 Wyoming passed a law requiring that civil engineers pass an examination in order to become registered to practice. Other states followed suit, and by 1920 18 states had passed similar laws and New York had instituted the engineer-in-training (EIT) certification. By 1947 all states had passed the law (Prasuhn 1995).

## Consulting Engineers

Consulting engineers have much influence in the civil engineering industry and provide employment for more graduates than any other sector of the profession. A number of nineteenth-century American engineers became famous for their work. For example, Dewitt Clinton prepared a report that led to the construction of the historic Croton Aqueduct in New York City. John A. Roebling designed the Brooklyn Bridge. These engineers practiced independently, were highly respected, and were the forerunners of today's consulting engineers.

Perhaps the oldest existing consulting engineering firm is Lockwood Greene Engineers Inc., of Spartanburg, South Carolina. The seeds of this firm were planted about 1832 by David Whitman of Rhode Island, who was generally regarded as the pioneer mill engineer in America. He was best known for designing the mills and water power in Lewiston, Maine. Another engineer, Amos Lockwood, assumed Whitman's work after he died in 1858. In 1882, Lockwood took on Stephen Greene as a partner, and Greene went on to build a premier textile engineering business. Although the firm expanded its ownership in the textile business, it suffered setbacks and engineering eventually emerged as the strongest part of the business. The firm was salvaged in 1928 as Lockwood Greene Engineers, Inc., and continues as a successful consulting engineering business (Lincoln 1960). Lockwood Greene has contributed an archive of historical drawings to the Smithsonian Institute's National Museum of American History.

The firm Stone & Webster, which was organized in 1889, grew from the friendship between two MIT freshmen. The vision of Charles A. Stone and Edwin W. Webster was to create "some sort of organization for undertaking electrical business in this country" (Keller 1989). About the 1890s, a number of other prominent firms had been established, including Parsons, Brinkerhoff, Quade & Douglas in New York; Sargent & Lundy in Chicago; Charles T. Main in Boston; and Burns & McDonnell in Kansas City, Missouri. Growth in the number of firms in the United States was rather slow until after World War II, increasing from 907 firms in 1940 to 4,943 by 1970 (Armstrong 1976).

## Construction Firms

Although most civil engineers do not actually work for contracting firms, they do work with contracting firms as consultants and construction managers. The Bechtel Corporation exemplifies the historical development of US contracting firms. Warren A. Bechtel founded the firm in 1898 in Oklahoma Territory with plans to focus on railroad grading. He then moved to northern California, where he pioneered the use of motorized trucks, tractors, and diesel-powered shovels in construction. His first diversification beyond railroad building was a contract from the U.S. Bureau of Public Roads in 1919 to build a highway along the Klamath River in northern California.

During the 1930s Bechtel's construction activities included large joint ventures and public works projects. The first was the Hoover Dam, built in 5 years, 2 years ahead of schedule. This was the largest civil engineering project undertaken to that time. From 1933 through 1936, Bechtel worked with the same partners to help build the San Francisco–Oakland Bay Bridge. In 1933, Stephen D. Bechtel, Sr. became president and expanded business to refineries and chemical plants and organized to support the war effort.

Bechtel has built some mega-projects, such as Turkey's Ankara-to-Gerede Motorway, which was built by Bechtel and the Turkish company Enka Insaat va Sanayi. The $1.6 billion project included a 115-km, six- to eight-lane toll road and a 114-km peripheral road around Ankara. Bechtel built San Francisco's Bay Area Rapid Transit system (BART), which at the time was the first fully automated and computer-controlled rapid transit system. In 1997 the American Society of Mechanical Engineers designated BART a National Historic Mechanical Engineering Landmark.

The Hoover Dam was Bechtel's greatest challenge in the early 1930s. At the time, it was the largest civil engineering project in the history of the United States at 726 ft high, 1,200 ft across the crest, and 660 ft thick at its base. Building the dam involved excavation of 3.7 million cubic yards of rock, erection of 45 million pounds of pipe and structural steel, and pouring of 4.4 million cubic yards of concrete.

Jubail Industrial City, a project for Saudi Arabia's Royal Commission for Jubail and Yanbu, is a metroplex on the Gulf of Arabia. The 360-square mile area includes a complete industrial and residential infrastructure, 16 primary industries, and a planned community. At its peak in the mid-1970s the work force reached 50,000. The total installed cost exceeded $40 billion.

Bechtel played a key role in putting out the Kuwait oil fires and in reconstruction after the Gulf War. They orchestrated the massive effort by mobilizing an international force of more than 16,000 workers to put out the 650 wellhead fires, stop the gushing flow of oil, and help resurrect the Kuwait oil fields. In addition, after Iraq released more than 11 million barrels of oil into the Arabian Gulf, Bechtel coordinated the clean-up.

On November 12, 1936, Bechtel and Bridge Builders, Inc. opened the San Francisco–Oakland Bay Bridge. At 8.5 miles, it remains one of the world's longest and busiest bridges, carrying more than 270,000 vehicles per day (Bechtel Corporation 1999).

# Engineering Failures

Part of our heritage includes failures, disasters, and catastrophes in engineered systems. These failures result from such natural and man-made threats and hazards as natural disasters, terrorism, bad designs and man-

**Table 2-1** Examples of engineering failure modes.

| System | Failure Modes |
|---|---|
| Transportation | Bridge failure |
| Water | Pipe break or drought |
| Environment | Toxic spill or contamination |
| Communications | Internet closure |
| Energy | Nuclear accident |
| Built environment | Terrorist attack |
| Intrasystem | Earthquake failure of networks |

agement, corruption, contamination, and wars and conflicts. Civil engineers play lead roles in protecting against all forms of disaster, including those from floods, earthquakes, hurricanes, mudslides, volcanoes, and terrorism. Increasingly, civil engineers are on the firing line in planning and designing against failures from these causes. With failures come experience, and by studying failures methods can be developed to avoid them in the future (Table 2-1).

In earlier years there was less recognition than there is today of the nature of threats, but there is much to learn from disasters. In Colorado, for example, the Big Thompson Flood Disaster of 1976, which killed nearly 150 people, taught lessons about flash flood hazards. The 1928 St. Francis Dam failure in California illustrated an early engineering failure that taught us lessons. Completed in 1926, the dam was to add to the water supply of Los Angeles, but it failed and resulted in the loss of at least 450 lives in the valleys of the San Francisquito and Santa Clara River basins. This disaster led to formation of the world's first dam safety agency; creation of uniform engineering criteria for compacted earthen materials; reassessment of all Los Angeles dams, including the Mullholland Dam; and formation of a process for arbitration of wrongful death suits (including processes that were used in the Loma Prieta earthquake in 1989). Subsequent investigations of the St. Francis Dam failure suggested that the failure was not due to piping but to placement of the left abutment on an ancient slide (Nunis 1995).

# Ethics

Although ethics is covered in depth in Chapter 12, it is important to touch briefly on the subject here. The evolution of the code of ethics was described by P. Arne Vesiland (Vesiland 1995) and is summarized here to place it into historical context. The Code of Hammurabi in Babylon stated: "If a builder has built a house for a man and has not made his work sound, and the house he has built has fallen down and so caused the death of the householder, that builder shall be put to death" (Martin and Schinzinger 1989).

During the Middle Ages, the codes of the guilds served as the de facto codes of ethics. When ASCE was organized, a code of ethics was deemed unnecessary; ethics were thought to be a matter of personal honor. It was not until 1914 that a code of ethics was adopted. It had six articles, which have been paraphrased by Vesiland as follows: Do not take bribes, do not speak poorly of colleagues, do not steal work, do not underbid a colleague, do not embarrass a colleague, and do not advertise. This code stayed unchanged for 20 years until a line was added in 1943 that forbids use of the advantages of a salaried position to compete unfairly with other engineers. In 1942 an eighth canon was added to prohibit acting in any manner that would bring discredit to the profession. In 1950 a canon prohibiting price competition turned out to be controversial. Also in 1950 the ASCE adopted another canon prepared by the Engineers Joint Council; this canon recognized for the first time that the primary goal of civil engineers was to serve the public good. This remained in place until 1962 when the board changed the preamble to recognize the public interest nature of civil engineering work. In 1972 the Department of Justice filed a suit claiming the code violated antitrust laws by restraining competition. The ASCE had to enter into a consent decree and remove the article from the code. The ASCE then added a footnote that explains that engineers are not prohibited from submitting a fee but must recognize that procurement of engineering services involves factors other than fees. In 1976, probably as a result of the Spiro Agnew affair, where the Vice President of the United States resigned under fire for actions he took while he was a Baltimore county executive, the ASCE abandoned its original code and adopted the Engineer's Council for Professional Development (ECPD) code, which emphasizes public welfare in stating that the primary responsibilities of engineers are public health, safety, and welfare. The debate over an environmental article continues.

## The Future

As was highlighted in Chapter 1, the future will be different for all workers, including civil engineers. This section outlines some of the projections and possibilities of the profession. Although the future cannot be accurately predicted, its possibilities can be explored. As the World Futures Society noted,

> The world changes so quickly it's hard to keep up. New inventions and innovations alter the way we live. People's values, attitudes, and beliefs are changing, and the pace of change keeps accelerating, making it difficult to prepare for tomorrow. By

**Figure 2-6**  A settlement on the moon.

studying the future, people can better anticipate what lies ahead. More importantly, they can actively decide how they will live in the future, by making choices today and realizing the consequences of their decisions. ... The process of change is inevitable; it's up to everyone to make sure that the change is constructive. (World Future Society 1999).

Trends in civil engineering were discussed in depth in Chapter 1, which outlined how civil engineering leaders see the future in general. A vision for the future of institutions is more complex. There needs to be an institutional model (including laws, incentives, behavior, customs, relationships, and organizations) that motivates technologic visions in order to create a positive future for humans and the environment.

Civil engineers have a great future, but they are not in it alone. Rather, civil engineers must work to apply technologies developed in all sectors to societal problems and learn how to involve the public in our work. Civil engineers can contribute to both technology and society in unique ways, and the work of civil engineers can stir the imagination (Fig. 2-6). Civil engineers are needed now and will continue to be needed in the future to provide structures and environmental living systems everywhere that people may choose to live.

# 3 Consequences of Civil Engineering

## Introduction

Chapter 2 discussed the dramatic impact civil engineers have had on society and the environment. Building on that discussion, this chapter examines the consequences of civil engineering and the roles of civil engineers in society. Civil engineers deal with the building and management of physical infrastructure systems, including roads and buildings, the water supply, and waste treatment. The consequences of civil engineering work can be bad or good, but they can never be ignored.

The economic prosperity of the United States can partly be attributed to the work of civil engineers in opening frontiers through the design and building of railroads and highway systems. To sustain development, engineers built water supply systems for irrigation and drinking water. Today, civil engineers are creating new systems to handle growth and urbanization throughout the nation.

Civil engineers have also played a more artistic role in society. Some buildings were built merely to satisfy someone's creative urge or ego. In the 1800s and early 1900s, for example, structures were built to be larger, taller, or grander than their predecessors. World fairs and expositions competed with each other, featuring such structures as the Ferris Wheel and the Eiffel Tower. A work of art on the scale of the Statue of Liberty required the participation of civil engineers.

The first section of this chapter explains how society cannot survive without having its basic needs met with the support of civil engineers.

Later sections describe how growth and development place demands on infrastructure and pressure on the environment. The final section explores the artistic side of civil engineering and traces the unintended consequences of civil engineering on the environment and on society. Also discussed are ways in which society promotes the positive consequences of development and seeks to rein in the negative consequences. In addition, the chapter provides suggestions about how civil engineers can create a better society while enjoying rewarding careers.

# Survival

Civil works have been necessary for the survival and success of society from ancient times, and they present unique opportunities and challenges to civil engineers today.

## *Shelter*

Shelter from the elements is a primary protection provided by civil engineers. People need homes to protect them from weather, floods, earthquakes, and other threats. Civil engineers are involved in community planning and construction development as well as in many other aspects of providing shelter. Civil engineers are active in such regulatory issues as developing codes and standards of practice for the building industry. The development of building codes leads to higher-quality design of all types of housing units and structures to keep people safe and secure. As knowledge grows, codes and standards are updated and people can be confident that structures built following those codes will provide safe shelter.

## *Security and Disaster Mitigation*

Survival involves preparing for and recovering from disasters. Disaster scenarios in which civil engineers have key roles include flood, earthquake, wind, drought, and even war or terrorism. For example, civil engineers have developed safer shelter from earthquakes. During the 1989 Loma Prieta and 1994 Northridge earthquakes in California, there were 60 and 61 deaths, respectively (Mileti 1999). Although all disaster-related deaths are unacceptable, it is important to note that tens of thousands of people died during the 1999 earthquakes in Turkey and Greece. Lower death rates in the United States result from strong building codes and strict enforcement of those codes.

Other examples of security and disaster scenarios in which civil engineering work has had a significant impact include the Dust Bowl of the 1930s; the 1993 Mississippi River floods; and Hurricane Mitch, which devastated Central America in 1999. In each case, civil engineers influenced the consequences of the events. For example, civil engineers helped design soil stabilization and drought control programs after the Dust Bowl. Likewise, they are part of the response and remediation teams for disasters such as floods and hurricanes.

## *Public Health*

Public health is an area that significantly impacts humanity . Among the most prominent members of the public health community, civil engineers have contributed a great deal to clean water, wastewater, air quality, hazardous waste management, and occupational health and safety. These activities all fall within the environmental component of civil engineering.

## *Clean Water*

Clean water is one of the most important resources required by people, but the quantity and quality of water often do not match the demands of society. Today, all parts of the world report water crises, whether from drought, contamination, or simply an imbalance between supply and demand. From early history, civil engineers have contributed to systems for providing water (see Chapter 4). Settlements usually developed near sources of water. Aqueducts were used in ancient societies, with the first incorporation of stone in them around 700 B.C. (Kirby et al 1956). Remnants of the aqueduct system of the Roman Empire can still be seen in Europe. As societies grew, the need to live in areas where water was scarce became a necessity. This required the transport of water through hydraulic engineering. New York City, for example, brings much of its water long distances from the Catskill Mountains.

In January 1801, Philadelphia became the first US city to develop a municipal water supply system. The city pumped water from the Schuylkill River via sluices and tunnels into 20,000-gallon wooden reservoirs. From the reservoirs, the water was initially conveyed through wooden conduits to public hydrants, households, and businesses (Koeppel 1994). Shortly afterward, New York City embarked on schemes to provide clean drinking water. Unfortunately, some of the schemes were political and aimed at money making rather than public benefit. In 1832, 3,500 New Yorkers died from cholera. By contrast, there were fewer than 900 cholera deaths in Philadelphia. In the 1834 cholera outbreak, there were again more deaths in New York City than in Philadelphia. The explanation for this difference was the use of clean water in Philadelphia and not in New York City.

New York City also experienced an uncontrollable fire in the early 1800s. The city finally mounted a major water supply project and built a canal system to bring water from the Croton River north of the city where it could be stored and distributed. Water first flowed into the reservoir in June 1842. This system continues in service today. Chief among the results following implementation of a water supply system in New York City was fire protection. A city previously ravaged by major fires now had a supply of water to combat this type of disaster. The number of deaths due to waterborne disease also decreased. Today New York City has an admirable water supply system.

Society faces greater challenges in providing water where it is not plentiful. Growth in the southwest United States led to an extensive system of dams, reservoirs, and canals to provide water for farms and cities. Much of the system was developed by the US Bureau of Reclamation, whose mission is to provide and manage water for 17 western states. Other parts of the system developed by state and municipal agencies enabled cities such as Las Vegas, Los Angeles, and Phoenix to become major metropolitan areas. The California Water Project is a state project with large-scale implications for the explosive growth in southern California of the past 30 years.

## *Wastewater Management*

Water can become contaminated through domestic use, farming, and industry. If not treated properly, water can threaten health and have a large economic impact. Collection of wastewater has been the job of society since ancient history: Brick-lined sewers dating to the eighth century B.C. were found near Nineveh. In London during the Middle Ages, sewer water was collected and deposited into the Thames River. Calculations showed where to discharge wastewater to avoid it returning to London during low tide. This approach eventually proved inadequate, and the modern approach is to treat contaminated water before discharge so that it does not degrade the environment. Wastewater can come from non-point as well as point sources, and many areas of human activity affect water quality. A great deal of environmental engineering work is aimed at managing wastewater from all sources.

## *Energy*

Energy is also necessary for survival. Civil engineers assist in developing fossil fuels, hydropower, nuclear energy, and renewable energy (e.g., from solar and wind sources). Civil engineers built small dams and facilities for early energy development from hydropower generated by mill wheels. Later, civil engineers built larger energy projects.

Energy is often misunderstood, and people complain about its environmental impacts at the same time that they increase their own use of it. The following joke helps explain the need to understand this concept: A professor was having a difficult time with students who had strong views on environmental protection. The professor said, "Well, if you don't like fossil power, don't like nuclear power, don't like hydropower, and don't even like wind, what do you like?" "That's easy," said a student, "just use electricity!"

## Growth and Development

After their survival needs have been met, people are next interested in improvement: building cities, creating businesses, and leading better lives. These improvements require the use of engineered systems to provide for social and economic development.

### *Infrastructure and Economic Development*

Chapter 10 describes how infrastructure and economic development relate to each other. Infrastructure issues are closely related to the interdependent economic–social–physical nature of urban systems. Built environments, such as cities or city–suburb complexes, are connected by networks that provide transportation and communications, are supplied by water and energy systems, and generate waste streams that must be processed. The driving forces behind built environments include growth, demographic changes, increased expectations, patterns of living, deteriorating systems, telecommunications, and increased complexity of systems. Society must invest its resources in transportation, water, energy, communications, the built environment, and waste management systems. Over the past two decades it became clear that increasing demands and past neglect created enormous capital investment needs, drawing society's resources away from other needs.

### *Transportation*

Transportation systems impact society, the economy, and the environment. Highways, railways, waterways, tunnels, and even space exploration are examples. For example, the interstate highway system, initiated in the 1950s, changed the face of America. Railroads impacted the United States in the nineteenth century and were the hottest stocks on Wall Street in the 1890s. Debates still rage about intercity and intracity transportation. Imagine the impact of the Chunnel, which connects the British Isles with the European continent. Waterways in the United States, starting with early canals such as the Erie Canal and continuing with waterways such as the Panama Canal and the Tennessee-Tombigbee, have had a great impact. Space exploration may be the final frontier of transportation, and civil engineers will design habitats for humans in space as well as pilot space vehicles.

**Figure 3-1**  Honshu-Shikoku Bridges.

An example of how transportation affects society is the Honshu-Shikoku Bridge system in Japan (Fig. 3-1). This tremendous array of bridges cost billions of dollars to build, employed a vast work force, and today connects two key islands of an advanced nation.

## *Energy*

In addition to survival, energy development enables settlement and industrialization. Civil engineers were part of modernization from steam engines, electrification, the gasoline engine, and nuclear energy. The results of these energy advances are the development of modern cities, large industrial complexes, and sprawl caused by low–cost energy policies and proliferation of automobiles. Today, the Internet and communications revolutions depend on the energy supply.

## *Water Engineering and Development of the American West*

Water resources engineering and development of the American West began after the Louisiana Purchase and included the explorations of Lewis and Clark, Zebulon Pike (who explored Colorado and for whom Pike's Peak is named), and the 1840s migration of the Mormons under the leadership of Brigham Young to the Great Basin of Utah. The Mormons built irrigation systems and showed how the land could be made productive. Later, as a result of the pressure of migration and gold mining, other parts of the West were developed, notably California and Colorado. These areas needed water supplies, for which the Colorado River was convenient. It was explored in the 1870s by Union Army veteran John Wesley Powell, who was a water development pioneer. Earlier, Lewis and Clark made notes about the water features and development possibilities of the West.

Western agriculture began to develop in the 1850s. Probably the biggest factor in water engineering in the West was the organization of the Bureau of Reclamation in 1903. Although the apparent reason for this organization was to build dams and water facilities, its main purpose was to provide a safe and secure basis for settlement of the West. These water developments have had a great impact, including conflicts between economic and environmental interest groups, water law, property rights, regional issues, and turf battles between government agencies (Grigg 1996). The general conflict is summarized in *Cadillac Desert: The American West and Its Disappearing Water*, which describes such western water issues as the Los Angeles water system (Reisner 1986). Today's urban sprawl in southern California and around Las Vegas provides a glimpse of the consequences of water development in the west.

## *The Aswan High Dam in Egypt*

Aswan High Dam in Egypt illustrates the dramatic consequences of civil engineering work. The Egyptian people settled for millennia in the valley and delta of the Nile River, and some 70 million people are sustained today in this narrow strip. They depended historically on the ebb and flow of the Nile, but a water storage project was needed to control the river's flow and provide for periods of drought. The Aswan High Dam was initiated in the 1950s with help from the Soviet Union. The political aspect of this dam construction illustrates great power politics after the 1956 Suez War. After the construction of the Aswan High Dam, the flood cycle was mitigated and abundant water was provided for irrigation development in the valley and delta. This enabled more population growth and provided the basis for agriculture that could not exist before the dam. Negative consequences of the dam will be discussed later.

## Art and Ego

Engineering work also has dramatic consequences in the world of art. This quote from Gustave Eiffel, developer of the Eiffel Tower, illustrates the point (Petroski 1997):

> Can one think that because we are engineers, beauty does not preoccupy us or that we do not try to build beautiful, as well as

solid and long lasting structures? Aren't the genuine functions of strength always in keeping with unwritten conditions of harmony? ... Besides, there is an attraction, a special charm in the colossal to which ordinary theories of art do not apply.

Eiffel was defending his tower against complaints that it was an affront to French art and good taste. It is a strong statement from the 1800s that continues to ring true today. As was pointed out by Eiffel, engineers—especially civil engineers—design the colossal structures (skyscrapers, dams, and bridges). There is beauty associated with the sheer size and magnitude of such structures.

## The Eiffel Tower and the Ferris Wheel

The Eiffel Tower is an example of the artistic nature of engineering structures. The story of the Eiffel Tower and its impact on the development of the Ferris Wheel illustrates how projects may be driven by the aspirations of nations and individuals (Petroski 1997). During the 1800s Britain and France held a series of expositions celebrating their histories. The Great Exhibition of 1851 in Britain resulted in the Crystal Palace (Billington 1983). France tried to outperform Britain through a series of expositions, ultimately resulting in construction of the Eiffel Tower in 1889. Not to be outdone, the United States planned the Colombian Exposition for 1893. One of the needs for this event was a structure that would display the ingenuity of US engineers. As Henry Petroski wrote, "American pride was at stake" (Petroski 1997). A structure to outdo the Eiffel Tower was required, and that structure was the Ferris Wheel.

## World's Tallest Building

Tall buildings, especially the tallest building, invoke pride, awe, and fascination. For many years the Empire State Building in New York City was the world's tallest building. Then the World Trade Center in New York City took on the title, followed by the Sears Tower in Chicago. During a 1980s television special called *Skyscraper*, New York developer Donald Trump stated that he wanted to build the world's tallest building in New York City because it bothered him that visitors to Chicago are greeted by a sign stating that Chicago is home to the world's tallest building. This type of provincial pride has now extended internationally. Currently the tallest building in the world, the Petronas Towers, is located in Malaysia. The desire to build the tallest continues to drive people to propose new structures. When this book was written, there were proposals for new buildings in São Paulo, Brazil, and Chicago that would stand taller than any existing structures. People have a need to build monuments to themselves and their communities. Civil engineering has responded by designing tall structures that must ultimately serve a useful function to society while satisfying the egos of the people funding the development.

## *Structural Art*

David Billington has shown us how civil engineering structures can be works of art. His thesis is that structural engineering has produced many beautiful creations with significant artistic value. Often, the unique use of such materials as reinforced concrete contributes to or drives the creative expression of the structural engineer/artist (Billington 1983).

Billington divides structural art into two periods: the Age of Iron and the New Age of Steel and Concrete. Examples from the Age of Iron include the Ferris Wheel and the Eiffel Tower. The New Age of Steel and Concrete is mainly associated with buildings and bridges. Robert Maillart designed concrete bridges with sleek members and beautiful curves that blended with the natural beauty of the surroundings. Heinz Isler designed concrete shell structures as roofs that are beautiful, thin structural works of art. Some of the world's tallest buildings reflect their internal structural design.

Engineering artistry is a very real and exciting aspect of the engineering profession. Engineers have designed many works of art that also serve society in functional ways. As a creative profession, civil engineering can contribute to society on multiple levels.

# Unintended Consequences

Many of the problems associated with the large-scale projects that typify civil engineering can be attributed to ignorance about unintended consequences. Unlike other engineering professions, civil engineers must operate in the public domain, where success and failure are apparent for all to see. Mechanical and electrical engineers often test designs in a laboratory and then let the market determine whether the design is a success (e.g., the minivan) or a failure (e.g., failed car designs). Civil engineering consequences often take on scales larger than the projects.

For example, the planning process of the Aswan High Dam involved great power politics. Would the United States or the Soviet Union help build the dam, thereby gaining influence in Egypt? After the 1956 Suez War, the Soviet Union seized the opportunity to replace the United States as the major power and offered to assist Egypt in the dam design, construction, and financing. Only in the early 1970s was the United States able to regain its influence with Egypt. During the 1950s and 1960s the Nile was dammed, and the project has been effective in regulating the river flow and has provided the people of Egypt with a reliable source of water. There have been side effects as a result of this project. Before the dam was built, annual flooding of the Nile River brought sediment and nutrients to farmlands in the delta and along the river. These sediments and nutrients are now captured behind the dam, creating a need for much more fertilizer application than before. Another consequence of dam construction is erosion and sedimentation. Clean water will erode more sediment than will sediment-laden water. Another consequence was a change in the Mediterranean River fishery that was present before the dam was constructed. A new fishery in the Aswan High Dam's reservoir, Lake Nasser, was created.

Evaporation loss is also a side effect of dams. In a desert, evaporation of 2 m of water annually from the surface of a reservoir amounts to a tremendous loss of scarce water. Another major effect of the Aswan Dam was an increase in salinity of the river delta at the point where the river enters the Mediterranean Sea. This increased saline has resulted in the destruction of a fishing industry in this delta and decreased agricultural production, resulting in loss of jobs. Although Egypt is still grappling with the consequences of the Aswan High Dam 40 years after its construction, most seem to agree that the country could not enjoy the quality of life it has today without the dam.

## *Interstate Highway System*

The interstate highway system in the United States had many intended and unintended consequences. Much of the motivation for the system evolved from the perception that it would provide for a more secure nation in terms of national defense. The ability to move manufactured goods throughout the country was considered a necessity for the defense of the country. This system also provided the ability to move consumer goods and food products more freely, resulting in a better-fed and more prosperous country. Unfortunately, the design also had some consequences that were not as positive. During the early development of the system, many designers anticipated that by increasing access to cities, rural living would become more attractive and stem migration to urban centers. Instead, more people migrated to urban centers for jobs and began to commute from the suburbs. This led to greater traffic congestion, one of the main problems resulting from growth and development.

## *Environmental Impacts: The Everglades*

Civil engineering projects often have significant environmental impacts, such as in the Everglades. Early in the twentieth century, parts of southern Florida were considered useless swamplands. The State of Florida became a pioneer in flood control, creating water management districts such as the SFWMD. The District covers all or part of 16 counties, with a population of over five million. SFWMD has a total area of 17,930 square miles, operates about 1,500 miles of canals and 215 primary water control structures, and has an operating budget approaching $200 million that is acquired primarily from property taxes. Its experience illustrates both positive and negative consequences of civil engineering work.

The two primary basins of the SFWMD are the Okeechobee Basin (based on the Kissimmee-Okeechobee-Everglades [KOE] ecosystem), and the Big Cypress Basin (which includes the Big Cypress National Preserve and some 10,000 islands). The environmental problems of the KOE ecosystem began early in the twentieth century, before Florida's explosive growth occurred. Agricultural drainage was viewed as a way to create productive land from "swamp." From 1913 to 1927, 440 miles of drainage canals were dug along the lower East Coast and south of Lake Okeechobee by the Everglades Drainage District. In 1926 and 1928 hurricanes in South Florida killed nearly 3,000 people living in farming areas. Damage was in the billions of dollars. The federal government responded with a levee around the lake. Then, from 1931 to 1945, several drought years brought low water supplies, peat fires, and saltwater intrusion. This was followed by a disastrous wet year in 1947 when over 100 inches of rain, twice the normal amount, hit the East Coast area. The response to two catastrophic hurricanes that year was the authorization of a Corps of Engineers flood control project and the creation of the Central and Southern Florida Flood Control District, which later became the SFWMD.

The first decade of the SFWMD focused on flood control, and enormous facilities were constructed, including dikes, pumping stations, canals, and levees. The second decade, 1959 to 1969, was a time of tuning up, testing the system, and continuing construction. Heavy rainfall in 1960 produced record lake levels in Lake Okeechobee, with a water level rise of 2 ft. The project proved that it could move large quantities of water and serve its purpose. Also in this period, providing water supplies for growing urban populations in South Florida became a priority, especially as the region faced the drought of 1961 to 1965. One of the largest projects of the 1960s was the channelization of the Kissimmee River. Over a 10-year period, the river was deepened and straightened by the Corps of Engineers.

These were all natural responses and projects for the conditions of that time. In the case of the KOE system, a river that had been 103 miles long was shortened to 56 miles, with a loss of about 35,000 acres of wetland habitat and a 90% reduction in migratory waterfowl. With the rise of environmental understanding in the 1970s, people began to call for work to remedy the mistakes of the past. By the mid-1980s, the District had studied alternatives for restoring the river, and the restoration project began in 1994. It is expected to take about 15 years to complete at a cost of some $372 million in 1992 dollars (Grigg 1996; SFWMD 1989). In 1994, Florida passed the Everglades Forever Act, which is intended to be a fundamental plank in the preservation and restoration program for the Everglades.

## Energy

Some unintended consequences of energy development include air pollution, strip mining, cooling water problems on fisheries, nuclear accidents, and global warming.

## Waste Management

Waste management has consequences that include ground water pollution, landfills, hazardous waste problems, dangerous hospital and toxic wastes, and nuclear waste. The acronym NIMBY (not in my backyard) applies to attempts to place any waste facility on a site where it will be noticed by practically anyone.

## Built Environment

The location of built facilities (e.g., shopping centers, office complexes, sports stadiums) has dramatic effects on living areas. Congestion, pollution, crime, and other social conflicts can occur if planning and development are not done responsibly.

# Promoting Positive Consequences of Civil Engineering Work

Civil engineers should seek ways to promote positive consequences. The quest for projects based on sustainable development, appropriate technology, and smart growth are just a few examples of attempts to promote positive outcomes.

## Government Incentives

Government incentives can promote such actions as environmental restoration and natural resources management. The idea is to provide positive incentives, rather than negative penalties, for desirable actions. The government provides many incentives for positive actions, such as best management practices, soil conservation, and set-asides for actions such as land restoration and cleanups. Figuring out what these incentives should be is a principal task of policy, often advised by civil engineers. Many of the incentives are in the tax code. Providing investment funds for infrastructure (e.g., issuance of tax-exempt bonds) is one way government can encourage positive action rather than simply penalizing negative actions.

Government incentives exist for growth and development patterns that meet societal needs, such as new cities and low-cost housing. Similarly, government can provide incentives for growth in ways that encourage positive things to happen. Two examples are low-cost housing, a big need in most communities, and tax increment financing for development in areas needing it. These are discussed further in Chapter 9.

## *Investments*

Investment in infrastructure is a way to stimulate growth, create jobs, and encourage employment and training programs. At the time of the New Deal, and even before, it was clear that multiple benefits could be achieved from investments in infrastructure by using infrastructure to stimulate jobs. Government investments in research and development can lead to sustainable development, as can the use of public funding resources to create investment pools (e.g., revolving funds and infrastructure development banks) so society can meet its basic needs (see Chapter 10). Investments came come from philanthropic sources rather than from government sources.

## *Public Education*

Public education is a way to promote positive consequences, particularly in topics such as environmental stewardship and citizenship in building communities.

## Controlling Negative Consequences

Preventing negative consequences is another way to approach the public good from civil engineering work and includes environmental regulation and emergency management. Environmental regulation is discussed in Chapter 11 and its impact on the economy in Chapter 10. The purpose of federal laws, such as the Clean Water Act and Endangered Species Act, is to prevent negative consequences. Unfortunately, the whole business of regulation is filled with conflict. Emergency management and public information tools can also be used to prevent negative consequences. Civil engineers are heavily involved in such programs (e.g., the national flood insurance program).

## Sustainable Development to Handle Consequences

Although sustainable development is discussed in detail in Chapter 10, it is important to outline some aspects here. Civil engineers have positive aspects on society, building the physical world that everyone depends on for transportation, clean water, and waste removal. Preventing the unintended consequences of our work also requires our attention.

The goal of sustainable development is to maintain the positive consequences of a project while minimizing the negative, unintended consequences. This requires long-term planning that often extends beyond traditional engineering. C.D. Johnson and R.M. Korol described principles for sustainable development (Johnson and Korol 1995):

- Anticipation and prevention: avoidance of environmental degradation at all stages
- Full-cost accounting (including environmental and social costs)
- Informed decision making and consideration of long-term planning and gains (including effective public participation)
- Living off the interest (the ability to treat natural resources as natural capital to be replaced as it is depleted [reused to recycled])
- Quality of development over quantity (placing the focus on durability, energy efficiency, and recycling)
- Respect for nature and the rights of future generations (including quality of life considerations that must include both the present and future)

These are excellent principles for engineering work. The need for informed decision making is of particular importance. Decision making is discussed in more detail in Chapter 6, and critical thinking is discussed in Chapter 7.

B.W. Baetz and Korol provide criteria for evaluating alternatives for sustainability: integration/synergy, simplicity, input/output characteristics, functionality, adaptability, diversity, and carrying capacity (Baetz and Korol 1995). These start by encouraging projects that integrate well with existing conditions, both natural and constructed. Simplicity is also encouraged as a method of accomplishing one's goals while mitigating unintended consequences. Civil engineers must incorporate concepts such as these in order to have a positive impact on society and minimize negative unintended consequences.

## Consequences on Civilization

Civil engineering work has tremendous consequences on civilization. Although much of the world is experiencing economic growth, other areas have been left behind (Fig. 3-2). Civil engineers are vital for both ends of this spectrum.

## Summary

Civil engineers are the designers, builders, and maintainers of the physical infrastructure that supports society. Civil engineers have made signif-

**Figure 3-2**   Hand construction of a canal in India. (Courtesy World Bank)

icant contributions to the health of people by providing clean water. Food production has also been greatly enhanced as a result of irrigation projects throughout the world. The world has become a more accessible place because of many transportation projects. Today, civil engineers are reaching into space with plans for establishing outposts based on the ideas of civil engineers working with other engineers and scientists.

Further, civil engineers caused some unintended consequences as a result of some of the projects they have undertaken. Because some these projects are very large undertakings, there have been many negative impacts. It is important to understand these consequences so that negative impacts can be minimized in the future through civil engineering leadership. Civil engineers must understand and use the concept of sustainable development to plan, design, build, and maintain the infrastructure for the future.

# 4 Work and Careers of Civil Engineers

## Introduction

The first three chapters of this book have discussed the heritage and consequences of civil engineering work. This chapter discusses the nature of the work itself. With careers and professions changing so fast, this is a critical subject that affects both the careers of individuals and the effectiveness of organizations.

Civil engineering work has many unique features and offers interesting and rewarding careers. Civil engineering combines technical work with management for economic progress, environmental management, and social welfare. Civil engineering also combines a variety of worthwhile activities into a people-serving profession that is financially rewarding. Although civil engineers do not always start at high salaries, more civil engineers advance into management than do other types of engineers. In addition, civil engineers manage complex and important societal systems.

This chapter presents a multidimensional picture of civil engineering work and careers. The information presented is designed to assist students in making informed choices about their programs and to give practitioners a guide for lifelong career paths. This chapter describes the following:

- Employment, technical areas, and functions of civil engineers
- Civil engineers as managers
- Characteristics of projects

- Professional background and education
- Registration and regulation of work
- Professional societies
- Lifelong learning and advancement
- Career tips
- Economic, social, and technologic trends facing civil engineers

## Engineering Work

Engineering work applies resources to create devices, structures, processes, and systems through the design process. Design and creation are at the heart of engineering work. Types of engineers vary by what is being designed or created (e.g., mechanical or electrical devices, chemical processes, infrastructure systems).

Engineering work starts when society has a need to be filled. To fill it, problems must be solved; devices, systems, and facilities developed; and management systems established. These tasks require land (resources), labor, and capital, all of which are central concerns of economics and all must be organized by work tasks. Knowledge and skills are divided into disciplines, occupations, crafts, and skills. Knowledge and work involve problem solving, analysis, design, production, and management, all common activities of engineers. Engineers must master bodies of knowledge organized according to discipline (civil engineering), occupation (consultant), industry (construction), skill or craft (design, modeling), or problem area (transportation).

As engineers join the community of workers, they are distinguished from related groups by the purpose and direction of the work, the technical sophistication, and the involvement in management (Table 4-1; Krick 1969; Beakley et al 1987). Essential characteristics of work that distinguish the related groups are science and mathematics, creative design, and management. Scientists and engineers use science and math more than do other groups. Engineers focus on creative design. Managers focus on organization and other tasks. The differences in the skill areas are not "either-or" but a matter of degree. The skill needed most uniquely by engineers is design, but engineers also need management skills. Although other disciplines (e.g., architecture) require design skills, the design skills needed vary among disciplines. These nuances can prove important when it comes to standards and selection of which discipline or occupation to undertake; the alternative is to be licensed for design and certification of different systems. Differences in and preparation for work determine qualifications to serve in the public interest and are professional and ethical issues.

The qualities of work that distinguish different types of engineers are content and context of knowledge needs. The content of the work of

**Table 4-1** The work of civil engineering compared to the work of other professions.

| Professional Group | Tools And Skills Required | Supervisory Roles |
|---|---|---|
| Craftsmen | Focus on tasks that require manual use of tools or equipment; skills often learned through practice and experience | Rare |
| Technicians | Perform routine equipment checks and maintenance; carry out plans and designs of engineers; set up scientific experiments | Seldom |
| Technologists | Apply engineering principles for production, construction, and operation; work with engineering design components | Occasionally |
| Managers | Oversee allocation and control of resources to create or operate something | Common |
| Scientists | Search for new knowledge or solution to science problems | Infrequently |
| Engineers | Identify and solve problems; innovate in applying science to produce feasible designs | Frequently |

engineers designing electrical devices is quite different from that of engineers designing buildings. Content differences in engineering work determine how engineers should be educated.

The context of work is also quite different. Devices may be designed within an industrial laboratory (which develops products for sale), but buildings may be designed in a consulting firm (which works for a public client who will pay from a public bond issue). Engineers must understand the cultures of these different contextual settings for their work. This relates closely to industries in which engineers work and determines the educational experience engineers require.

# Trends in Engineering Careers

Engineering careers are experiencing definite trends. These trends have been identified over the past three decades (Burton et al 1998):

- There are two to three times as many engineers today as there were 30 years ago.
- Younger engineers have more and higher degrees than do older engineers.
- Engineers spend more time on management and human resource activities than they used to.

**Table 4-2** Order of work activity for engineers.

| Activity | Percent Involvement |
|---|---|
| Design | 66 |
| Computer applications | 58 |
| Management | 49 |
| Development | 47 |
| Accounting and other financials | 42 |
| Applied research | 39 |
| Quality or productivity | 33 |
| Employee relations | 23 |
| Sales | 20 |
| Basic research | 15 |
| Production | 14 |
| Professional services | 10 |
| Other work activities | 8 |
| Teaching | 8 |

- Electrical engineering is the predominant field, replacing mechanical engineering; and the field of software or computer engineering is growing rapidly.
- Design has replaced development as the engineer's main job function.
- The work force is a little more diverse.
- Engineers receive extensive continuing education and training.

Design tops the rank order of work activity for all engineers (Table 4-2). A further breakdown of the data shows that mechanical engineers do more design, electrical engineers work more with computer applications, and civil engineers are more involved in management (Burton et al 1998).

## Civil Engineer Employment, Technical Areas, and Functions

There are three defining characteristics in the domain of civil engineering activities: the employer, the technical area, and the general function (Fig. 4-1).

### Civil Engineering Employers

Civil engineers work on six large system (industry) types: transportation, water, waste management and environment, energy, communications, and the built environment. These six industries shape the content and context of civil engineering work, although the lines between them are often blurred and they are not necessarily all-inclusive. In these indus-

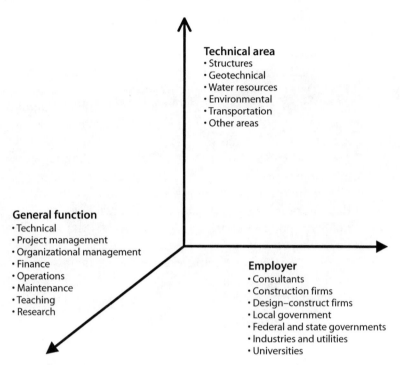

**Figure 4-1**   Civil engineering activities.

tries, civil engineers work with structures, systems, processes, and devices, focusing less on inventing new ones than on assembling components already developed. There are always exceptions, and the structures, systems, and processes that civil engineers develop are normally unique rather than mass produced. This has important implications for civil engineering design, as opposed to other disciplines. Civil engineers will often find design work in uncharted water, outside of standard practices. This will require higher relative costs for design than a project that allows for amortization of design costs over many copies of the original product.

Civil engineers work more in the public sector than do other engineers. Civil engineers work with public officials, politicians, laws and regulations, and public involvement programs. Their involvement in public-sector decision making exposes them to such civil engineering enigmas as NIMBY, BANANA (build absolutely nothing at or near anyone), NOPE (not on planet Earth), and NOTE (not over there either). Their focus on work in the public arena makes civil engineers more reliant on topics related to government and public involvement than are workers in other disciplines and makes them ready for the CAVE (citizens against virtually everything) phenomenon.

Highly technical industries often get more attention than does engineering. According to the National Academy of Engineering (NAE), for example, the most admired engineering achievements from 1964 to 1989 were the moon landing, application satellites, microprocessors, computer-aided design (CAD) and management (CAM), computer-assisted tomography (CAT scan), advanced composite materials, the jumbo jet, lasers, fiber-optic communication, and genetically engineered products (NAE 1989). Most of these engineering feats have civil engineering components.

In 2000 the NAE selected the top 20 achievements of the past century, and this time civil engineering scored better. Several of the greatest achievements related directly to infrastructure, particularly electrification, automobiles, airplanes, safe and abundant water, electronics, radio and television, agricultural mechanization, computers, telephone, air-conditioning and refrigeration, interstate highways, space exploration, the Internet, imaging technologies, household appliances, health technologies, petroleum and gas technologies, laser and fiber-optics, nuclear technologies, and high-performance materials (NAE 2000).

Civil engineering achievements, being mainly one of a kind, are also highly significant. The ASCE listed the top 10 achievements of the twentieth century as follows (ASCE 2000):

- Airport design and development
- Dams
- Interstate highway system
- Long-span bridges
- Rail transport
- Sanitary landfills and solid waste disposal
- Skyscrapers
- Wastewater disposal
- Water supply and distribution
- Water transportation

The emphasis on high-technology achievements underscores the point made in Chapter 1 that civil engineers work on problems that are not always high profile, technologically speaking, but that civil engineering projects should apply new technologies.

Civil engineering employers have been categorized by the industry–occupation matrix (Table 4-3), which cross-tabulates employment by occupation and industry (Ellis 1997). Some 183,102 civil engineers were employed in 1996. The largest employer was consulting (engineering services combined with management and accounting services) at 81,340, followed by state and local governments at 58,653, for a total of 139,993 jobs (76% of total civil engineering employment). Next in order were the federal government (12,622); construction (10,852); all manufacturers (8,236); and electric, gas, and communications utilities (combined; 3,905).

The distribution of civil engineering employment does not change rapidly. The ASCE surveyed its membership in 1973, with 40,000 members responding (Armstrong 1976). The largest faction of respondents was consulting engineers (32.1%), followed by federal government

## Table 4-3  The industry–occupation matrix (Ellis 1997).

|  | Occupation | | | | | |
|---|---|---|---|---|---|---|
| Industry | Chemical Engineers | Civil Engineers | Electrical Engineers | Mechanical Engineers | All Others | Managers* |
| Total engineers and managers | 48,126 | 183,102 | 354,079 | 223,846 | 315,690 | 342,893 |
| All manufacturers | 32,916 | 8,236 | 164,207 | 132,872 | 165,628 | 133,233 |
| Nonmanufacturers (except government) | 13,780 | 103,591 | 154,940 | 79,129 | 110,045 | 178,732 |
| Communications utilities | 39 | 993 | 18,132 | 361 | 4,339 | 8,480 |
| Construction | 1,235 | 10,852 | 9,429 | 5,093 | 2,837 | 3,282 |
| Electric and gas utilities | 605 | 2,912 | 15,620 | 3,179 | 6,295 | 7,437 |
| Engineering services | 5,608 | 74,900 | 35,630 | 29,839 | 18,498 | 35,049 |
| Management services | 1,726 | 6,440 | 7,421 | 3,848 | 11,804 | 12,443 |
| Research and testing | 3,108 | 2,119 | 20,039 | 8,215 | 17,008 | 18,004 |
| Other services and utilities | 529 | 3,190 | 6,796 | 4,394 | 11,090 | 14,736 |
| Federal government | 1,149 | 12,622 | 29,095 | 10,566 | 26,322 | 13,326 |
| State and local government | 281 | 58,653 | 5,837 | 1,279 | 13,694 | 17,602 |

*Managers refers to engineering, science, and computer managers.

employees (15.5%). State government was next at 10.2% (adding state highways to other state government), followed by construction at 9.4% and county or municipal government at 8.1%. Industry came in at 7.2%, and educators (despite their prominent roles in the ASCE) comprised only 6% of the work force. The anomaly is the lower percentage of local government employees, reflecting the fact that much engineering work has "devolved" to local government over the past two or three decades.

Most consultants can be found in firms that belong to the ACEC, which lists approximately 5,700 independent member firms employing over 250,000 people in all occupations. Of the member firms, 75% have fewer than 25 employees (ACEC 1999). There are about 17,000 consulting engineering firms in the United States (US Department of Commerce 1998). Of the 500 largest design firms, total revenues can be computed to be on the order of $40 billion (ENR 1996). Larger firms dominate revenues and employment, but smaller firms add significantly to total revenue, making engineering services over a $50 billion industry based on revenues.

State government primarily employs engineers in transportation departments and regulatory agencies. Local government engineers work mainly in public works and utility organizations, with the following functional areas:

- Municipal
- Streets and transportation

- Facilities
- Drainage
- Water supply
- Wastewater
- Planning and community development
- Solid waste
- Geographic information systems (GIS), surveying, and mapping

A generation ago, the profile of a civil engineer would have been like the dictionary definition: "an engineer who designs and constructs public works." The work of a civil engineer today is much different. The following list (which corresponds to lines in the industry–occupation matrix) highlights the work of civil engineers:

- Medium-level engineer in a consulting firm
- Public works or utility manager in a medium-sized city
- Corps of Engineers employee in a district office
- Regional engineer for a state transportation department
- Regulator in a state environmental agency
- Construction manager in a medium-sized construction firm
- Product specialist in a concrete pipe company
- Facilities development engineer for a corporation
- Software developer for a small, specialized firm
- Civil engineering professor at a state university

## Technical Areas of Civil Engineering

The ASCE technical division enrollments are a rough measure of the proportion of jobs available in each category (Table 4-4). Engineers can register for more than one technical area, so the numbers are somewhat inflated.

**Construction Engineering.** Construction engineers work in one way or another with construction—building construction, heavy construction, utilities, and specialty contracting. Civil engineers are not always employed by contractors: The industry–occupation matrix shows that only 6% of civil engineers work for contractors. Civil engineers are involved with planning, scheduling, and construction processes. The American Council for Construction Education maintains information about accredited construction education programs (American Council for Construction Education 1999). There were 43 programs in 1999, including schools of construction within engineering colleges, departments of building science within architecture colleges, construction management programs either within or outside of engineering schools, and departments that educate technologists. There are various routes to becoming involved in construction management, and engineering is only one of them. Figure 4-2 illustrates construction in progress.

**Table 4-4**   **American Society of Civil Engineers (ASCE) Division Enrollments (ASCE 1997, 1998).**

| Division | 1997 | 1998 |
|---|---|---|
| Aerospace | 2,029 | 1,792 |
| Air transport | 3,751 | 3,353 |
| Architectural engineering | 4,332 | 5,604 |
| Codes and standards | 11,954 | 11,622 |
| Cold regions engineering | 2,944 | 2,735 |
| Computer practices | 18,654 | 17,611 |
| Construction | 37,998 | 35,373 |
| Energy | 6,683 | 5,835 |
| Engineering management | 30,815 | 28,462 |
| Engineering mechanics | 7,197 | 6,605 |
| Engineering school faculty | 4,172 | 4,172 |
| Environmental engineering | 31,256 | 28,330 |
| Forensic engineering | 7,856 | 7,743 |
| Geomatics | 11,008 | 10,158 |
| Geotechnical engineering* | 22,549 | 7,252 |
| Government engineers | 3,452 | 4,301 |
| Highway | 23,153 | 21,499 |
| Lifeline earthquake engineering | 6,407 | 6,011 |
| Materials engineering | 7,476 | 7,147 |
| Pipeline | 8,843 | 7,371 |
| Structural* | 32,001 | 14,454 |
| Urban planning and development | 15,099 | 14,031 |
| Urban transportation | 13,133 | 12,038 |
| Water resources engineering | 26,090 | 23,699 |
| Water resources planning and management | 20,082 | 18,152 |
| Waterway, port, coastal, and ocean | 8,923 | 8,266 |

*The ASCE initiated the Geotechnical and Structural Institutes in 1998.

**Environmental Engineering.** Environmental engineers initially worked with water and wastewater, waste management, and air quality. Now their work includes such ecologic issues as endangered species. The mission statement of the American Academy of Environmental Engineers (AAEE) states that members should be "dedicated to excellence in the practice of environmental engineering to ensure the public health, safety, and welfare to enable humankind to co-exist in harmony with nature" (AAEE 1999). The AAEE mission statement goes on to explain the history of environmental engineering (AAEE 1999):

> Environmental engineering, or sanitary engineering as it was called before 1970, began in the United States in the 1830s with the design of water supply systems. The industrialization and

**Figure 4-2**  Steelwork at Coors Field (Denver, Colorado). (Courtesy L.P.R. Construction Co., Loveland, Colo; photograph by James Digby)

urbanization following the Civil War created life-threatening water and air quality problems that occupied the fledgling profession in the late 1800s and into the 20th century. By the onset of World War II, safe drinking water was the norm throughout the United States. Continued industrialization during and after World War II radically increased all forms of environmental pollution. But, the accompanying economic boom also unleashed pollution control technologies and gave rise to the experts who applied and advanced them.

The AAEE mission statement further explains that environmental engineering became a multifaceted field, representing public health, engineering education, civil engineers, water works, and the water environment federation. Later, other associations joined in as well, representing chemical engineers, public works, professional engineers, environmental engineering professors, mechanical engineers, and solid waste.

**Geotechnical Engineering.** Geotechnical engineers work on foundations, tunnels, retaining walls, liners, and related problems and are increasingly involved in environmental matters. Originally, the discipline was called *soil mechanics*. Recently, ASCE created the Geo-Institute (GEI) from its Geotechnical Division (ASCE 1999). In addition, geotechnical engineers have a presence in these associations: the American Geological Institute, American Rock Mechanics Association, Association of Engineering Geologists, Canadian Geotechnical Society, Deep Foundations Institute, Environmental and Engineering Geophysical Society, Geosynthetic Institute, International Association of Engineering Geology and the Environment, Industrial Fabrics Association International, International Association of Foundation Drilling, International Erosion Control Association, International Geosynthetic Society, Geo-Council of the National Council for Geo-Engineering and Construction, International Society for Soil Mechanics and Geotechnical Engineering, and other professional firms practicing in the geosciences.

**Structural Engineering.** Structural engineers design buildings, bridges, offshore structures, and other facilities. Like geotechnical engineers, structural engineers formed their own institute within the ASCE, the Structural Engineering Institute (SEI) (ASCE 1999). Like the GEI, the SEI demonstrates the span of its interests by its associations, which include the American Concrete Institute, Council of American Structural Engineers, FEMA, Housing and Urban Development (HUD), National Institute of Standards and Technology, codes and standards organizations such as the American National Standards Institute and the Building Officials Code Administrators, American Institute of Steel Construction, American Plywood Association, American Society for Testing and Materials, National Association of Home Builders, National Concrete Masonry Association, and National Institute of Building Sciences.

**Transportation Engineering.** Transportation engineers comprise the broadest categories under ASCE's technical tent. Transportation engineers work with roads, highways, bridges, urban transit, railroads, aviation, aerospace, and water transportation. Civil engineers play a central role in planning, design, construction, and operations of all these transportation systems. Today, the rise of intelligent transportation systems offers new opportunities to apply operations management and information technology to solution of problems. As a result of large expenditures on transportation systems, much civil engineering employment will be related to transportation.

**Water Resources Engineering.** Water resources comprises another large part of civil engineering work and includes planning and management of water resources systems, flood control, irrigation systems, canals, pipelines, and dams. Water resources engineering overlaps with environmental engineering. With increasing recognition of the interdisciplinary character of water issues, water resources engineering has become very diverse. A variety of crises have focused attention on additional dimensions of water resources, including climatic shifts, rapid sociopolitical changes, transboundary dependencies, increased hazards, and challenges from rapid technologic developments. We recently studied water issues and concluded that the most pressing are release of water for new and expanding demands, habitat management for fish and wildlife, maintaining productive agriculture, expanding economies dependent on water, security against floods and droughts, minimizing

72 ♦ *Civil Engineering Practice in the Twenty-First Century*

**Figure 4-3**  A water resources engineer inspects an irrigation canal in the western United States.

water quality degradation, and developing appropriate institutions for a complex and interdependent environment.

Figure 4-3 depicts a water resources engineer inspecting an irrigation canal in the western United States. In dry regions of the world, agriculture depends largely on irrigation systems.

**Other Specialty Areas.** Other specialty areas in civil engineering include ocean engineering, facilities management, wind engineering, materials engineering, urban planning, and systems engineering. Allied fields also attract civil engineers, involving such areas as real estate and community development, infrastructure and public works management, and city management. All these arenas create a broad work environment and opportunities for civil engineers.

## Functions of Civil Engineers

The functions of civil engineers are widely varied. For example, the ASCE consulting engineering manual lists consultation and advice, feasibility studies, field studies, environmental studies, reports, cost estimates, design drawings and documents, testing, construction administration, financial studies, expert testimony, and preparation of operations manuals as possible tasks for civil engineers (ASCE 1996). Table 4-5 lists a taxonomy of civil engineering work, in rough order by frequency of work.

## Management Activities of Civil Engineers

Management is a special function of civil engineers that is not as well understood as are other categories of work (Table 4-6).

## Characteristics of Civil Engineering Work

Chapter 1 outlined how civil engineering work involves public sector spending and regulatory issues, involves private practice more than do

**Table 4-5** Taxonomy of civil engineering work.

| Type Of Work | Examples |
| --- | --- |
| Design of facilities and systems | Design drawings and documents |
| Project management | Project planning and oversight |
| Environmental studies | Environmental evaluations with other disciplines |
| Communications | Reports and public presentations |
| Construction administration | Construction inspection and supervision |
| Cost estimates | Cost estimates for projects and systems |
| Regulatory work | Permit studies and administration |
| Facility management and operation | Operations manuals |
| Financial studies | Project financial analysis |
| Organizational management | Supervision of personnel and teamwork |
| Project evaluations | Planning and feasibility studies |
| Use of computers and information technologies | Modeling and database work |
| Business development and marketing | Opening of new engineering markets |
| Data collection and analysis | Field studies of various kinds |
| Legal support | Expert testimony |
| Research | Inquiries of different types |
| Testing | Laboratory analyses |
| Advisory services | Consultations and advice |

**Table 4-6**  **Management activities of civil engineers.**

| Management Activites | Examples |
|---|---|
| Project management | As in construction, focus on a particular system using resources, people, team members, contracts |
| Business management | All aspects of private business management, including consulting firms |
| Public organizational management | Public situations, such as city or agency management, public works management |
| Operations management | Applying management to ongoing situations and systems, including traffic controls, water treatment plants |
| Human resources management | Administration of supervisory systems |
| Environmental regulation | Focus on decision making about how to regulate in environmental arena |
| Program management | Developing and offering a program, such as an outreach program or service operation of some kind |

other engineering disciplines, attracts more interest in professionalism, has more influence on the construction and infrastructure industries and on environmental regulation than do other engineering disciplines, has a larger social component than do other engineering disciplines, and is more stable than are other engineering disciplines.

## *Civil Engineering Work Is Highly Varied*

The dictionary definition of civil engineering is too limited—civil engineers do much more than design and construct. Civil engineers are involved in many types and phases of work (e.g., design, construction, project management, research, system management).

## *Civil Engineering Projects Are Unique*

Civil engineers often work with customized, site-specific designs that result in one-of-a-kind projects, facilities, and systems. Members of other engineering disciplines work more on mass-produced products, whereas civil engineering projects have a low design intensity factor (which measures the cost of design and development divided by the cost of one item). Because the cost of engineering might be approximately 8% to 12% of a project, the index is often 0.08% to 0.12% for one-of-a-kind projects, whereas the index is much greater for mass-produced products that are designed only once (e.g., commercial aircraft, automobile models, computer chips).

### Civil Engineering Work Is in the Public Domain or in Environmental Protection

Civil engineers often create or work with facilities that are owned by communities and/or the government. This is one reason why civil engineering work is closely identified with public works. Likewise, a primary customer of civil engineers is the environment, both local and general.

### Civil Engineering Work Is People-Oriented

Civil engineers commonly interact directly with owners and user communities. This creates a need for skills in management and public participation (see Chapter 8). Civil engineers are not always working at the computer; most also work a great deal with the public.

### Civil Engineering Projects Have Long Lives

The projects and facilities that civil engineers create often have very long lives, outliving their designers and builders and becoming key features (sometimes landmarks) of the physical community. Examples of this can be seen in large bridges, office buildings, and even water treatment plants.

## Examples of Civil Engineering Projects

To illustrate the range of civil engineering, this section describes three projects: a midsized office building, a water treatment plant, and an upgrade to a major urban arterial. These projects illustrate the myriad issues that civil engineers must face. Although they are primarily design related, these projects require substantial skill and experience. These projects provide an idea of the variety available in civil engineering work, and there are many more examples.

### Midsized Office Building

The feasibility study for a midsized office building built by a private developer in a suburban area must address utilization, size, and financing. Land acquisition must consider zoning and other land use issues. Selecting design professionals and securing financing are management activities for the owner. The project approach (e.g., design and bid, design-build) must be determined. The design process will include consideration of the site, parking needs, drainage, architecture layout, foundation and other geotechnical aspects, hydrology/drainage, transportation, parking, access to street/highway, structure, and more. All codes and standards must be followed. Phases will include design, construction, operation and maintenance, remodeling, expansion, and possibly eventual demolition of the project.

## *Water Treatment Plant*

Creation of a water treatment plant for a city of 30,000 involves system and community aspects. In this case, engineers need studies of population projections, long-term plans, source and amount of water, and other important data. Preliminary planning will show how the treatment plant fits with existing systems such as other plants or major supply lines. Also, decisions must be made regarding how much to build now versus later. An environmental impact statement (EIS) and various permits may be required. Financing may include state or federal participation and involve bonds, elections, taxes, or other public issues. Design will bring all the physical and construction factors together. Construction will include such issues as how to tie water mains into the system and plan for minimal disruption of service. Operations must include a trial operational period to shake down the facility before full operation and might include partial operation in the initial years before the full design demand is felt. Modifications and/or expansions may be needed to meet future demands.

## *Upgrade to an Urban Arterial*

Upgrade of a major urban arterial street or freeway involves planning; financing; evaluation of alternatives; and studies of growth projections, traffic records, and system characteristics. The compatibility of the project with urban growth plans and zoning must be evaluated. Input from the political process, including both state and local governments, is essential. Alternatives might include a mix of streets, rapid transit, buses, and high-occupancy vehicle (HOV) lanes. Consideration of air quality and environmental concerns occurs during planning. A systems evaluation of potential impacts on adjacent parts of the transportation system (e.g., local streets in areas near interchanges and exits) will be required. The EIS and permit process might involve public hearings. Design will bring in several components, such as freeway geometry, pavement design, bridges, signage, and displaced utilities. Transportation requires attention to construction with minimal impact and trading of construction time and project costs. During construction, it will be critical to keep the freeway open and minimize adverse impacts on traffic, utilities, and other systems in the corridor. The project may involve more than one prime contractor. Bringing materials to the site may require planning of night work, truck restrictions, and noise restrictions. Inspection, acceptance, signage, lane painting, and traffic management will be required during the opening phase. Monitoring of the operation will show how well it is working.

# Requirements and Regulation of Civil Engineering

Engineering practice is regulated by ABET, codes and standards, and registration and control of professional engineers (handled by state registration boards).

## Regulation of Engineering Education by the Accreditation Board of Engineering and Technology

The background and professional requirements required to become a civil engineer reflect the KSAs that civil engineers must possess. KSAs include such tasks as problem identification, planning, decision making, design, construction, operation, and capital management as well as others that relate to areas of specialization. Civil engineers are no longer spending as much time on traditional tasks and skills such as surveying and hand calculations as they are on new skill areas such as computer simulation modeling, writing, and management of interdisciplinary design teams.

Regulation of engineering education seeks to keep up with these changes and provide an educational system that responds to current needs. There are approximately 250 four-year civil engineering programs in the United States, all regulated by ABET. ABET is a nongovernmental organization that provides a peer review process to ensure quality in engineering and technology education. ABET also provides for institutional accreditation of civil engineering programs at 2,300 sites around the country located at over 500 colleges and universities, including engineering technology and engineering-related programs. ABET is recognized by the US Department of Education and the Council for Higher Education Accreditation for responsibility in its areas of expertise.

The ABET is a federation of 28 professional engineering and technical societies, of which ASCE is one. ABET manages the voluntary process that institutions choose to undertake. The organization also assists with important decision-making activities by such groups as students choosing an education program, parents seeking the assurance of a quality education, institutions seeking to improve educational programs, employers recruiting well-prepared graduates, state registration boards screening applicants for entry into practice, and industry representatives seeking to voice educational needs to institutions.

## Regulation of Engineering Designs and Activities by Codes and Standards

As outlined in Chapter 11, there are many legal instruments and regulations that govern the practice of engineering by establishing constraints and guidelines. There are also codes and standards, which are extremely important in civil engineering because public safety and health as well as environmental quality are involved. Codes and standards are self-regulatory (voluntary rules and guidelines). If an industry does not establish codes and standards, then government may step in and do so.

For example, the ASCE has numerous codes and standards. These include codes and standards for transportation systems, electrical transmission, gas shutoff systems for earthquake protection, nuclear facilities, specification formats, microtunneling construction, residential structures on expansive soils, shore protection, concrete box culverts, independent peer review of projects, inspection of dams, pile foundations, subdivision plans, artificial recharge of groundwater, atmospheric water management, oxygen transfer, border water quality, urban drainage, and water regulations.

In addition to the ASCE, other organizations also promulgate and coordinate codes and standards. For example, the SEI has codes and standards for air-supported structures, design loads during construction, engineered wood construction, fiberglass-reinforced plastic stacks, flood-resistant construction, masonry, building design loads, seismic isolation systems, stainless-steel cold-formed sections, steel decks with concrete, steel cables in buildings, fiber composites and plastics, structural condition assessment and rehabilitation, structural steel beams, tensioned fabric structures, and wind tunnel testing. Ethics can also play a significant role in regulating civil engineering practice (see Chapter 12).

## *Regulation of Professional Practice*

State registration boards, which exist in every US state, are coordinated by the National Council of Examiners for Engineering and Surveying (NCEES). The general purpose of the NCEES is to provide leadership and professional licensure of engineers and land surveyors through excellence and uniform laws; licensing standards; and professional ethics for the protection of the public health, safety, and welfare. They are also charged with shaping the future of professional licensure. The legal aspects of the practice of engineering are complex and are summarized in Chapter 11.

# Civil Engineering Education

Recruitment of civil engineers into the profession begins long before the college level. Students who exhibit skills in mathematics and science become identified as potential engineers and scientists early in their educational careers. By the time these students reach high school, they have already decided on a career path and normally apply to an accredited engineering college and enter into an electrical, mechanical, civil, or chemical engineering program. They might choose smaller, niche fields such as nuclear, industrial, or agricultural engineering. Surprisingly, many practicing civil engineers did not go through the formal civil engineering educational process but arrived at their career destinations via studies in other majors.

As technology becomes more complex, it is increasingly necessary to have more than a minimum 4-year degree. In recent years the ASCE has been considering a 5-year degree as a minimum entry to the practice of civil engineering. This is a controversial subject because the same goal can be achieved with a 4-year degree and another year of graduate study leading to the master's degree. Master's degrees are very useful to civil engineers and other professionals who seek to practice in specialized areas. After obtaining a graduate degree, engineers must still engage in lifelong learning. Increasingly, state registration boards are requiring certification of continuing education credits to renew registration. Many resources are available for continuing education, both self-paced and in organized ways (e.g., university courses; short courses offered by professional associations; self-paced learning on the Internet; review of journal articles, books and other materials; teaching courses for others).

## Curriculum Requirements

Colleges are accredited under ABET criteria, which seeks to ensure that civil engineering curricula are planned with the advice of constituent groups and that outcomes are measured (ABET 1999). ABET recommends 11 required skill areas (encompassing design, breadth, experimentation, and practice), and the ASCE added to that list (ASCE 1999):

> a major design experience based on the knowledge and skills acquired in earlier course work and incorporating engineering standards and realistic constraints that include most of the following considerations: economic; environmental; sustainability; manufacturability; ethical; health and safety; social; political; proficiency in a minimum of four recognized major civil engineering areas; the ability to conduct laboratory experiments and to critically analyze and interpret data in more than one of the recognized major civil engineering areas; the ability to perform civil engineering design by means of design experiences integrated throughout the professional component of the curriculum; and an understanding of professional practice issues such as procurement of work, bidding versus quality based selection processes, how the design professionals and the construction professions interact to construct a project, the importance of professional licensure and continuing education, and/or other professional practice issues.

Chapter 1 of this book listed the usual university requirements for an educated person, expressed by the all-university core curriculum. What a college graduate needs to know and what an engineer needs to know are similar in many ways, but there are differences in level and context (Table 4-7).

In addition to the areas shown in Table 4-7, the ASCE requires basic engineering, more problem solving, at least four technical areas, experiments and data, engineering tools, design process, engineering standards, knowledge of issue areas (economic, environmental, sustainability, manufacturing, ethics, health and safety, social, and political), and professional practice.

**Table 4-7** Educational skill level comparison.

| All-University Core Curriculum | Purpose | Additional Engineering Requirements |
|---|---|---|
| Freshman seminar | Provide an integrative experience for first-year students | |
| Written and oral communication | Establish functional skills in communication, possibly including a second language | |
| Mathematics | Basic skills in computing | More mathematics |
| Logic/critical thinking | Foundation for analysis and problem solving | |
| Biological/physical sciences | Basic knowledge in science areas | More physical sciences |
| Arts/humanities | At least a basic appreciation of arts and humanities | |
| Social/behavioral sciences | Foundation knowledge of the social and behavioral sciences | |
| Historical perspectives | A sense of the past | |
| Global and cultural awareness | Awareness of current context and events | Contemporary issues relating to engineering practice |
| US public values and institutions | Knowledge of how US society and government evolved | Civil engineers must understand public involvement |
| Health and wellness | Appreciation of personal health and wellness | |
| Integrating competencies: writing, speaking, and problem solving | Demonstrating the ability to apply communication and thinking skills to solve problems | Engineers must communicate and solve problems in a wide span of contextual situations |
| Building on foundations and perspectives | Demonstrating the ability to apply knowledge to solve problems | |
| Capstone course | Final integration of knowledge and skills | Complex design experience and function on multidisciplinary teams |

Educators are required to address these issues in the undergraduate curriculum. At the graduate level, the focus is on technical specialty areas. Continuing education and lifelong learning courses cover all areas, providing professional development hours for registration certification. Many other questions can be formulated about educating civil engineers, including the following:

- Should the master's degree be the entry point to professional practice?
- How much concentration should there be on management?
- What is the societal character of civil engineering?
- How are future leaders attracted to civil engineering?
- How are environmental issues included in educational programs?

## Professional Engineering Societies

Professional societies have a significant role in shaping the work and careers of civil engineers. Because it is the main professional society for civil engineers, work force issues are one of the main concerns for the ASCE. As the civil engineering profession evolves, it must continue to change as it has in its first 200 years to meet the increasingly complex physical needs of society (Wisely 1974; Watson 1988). Success in developing and managing infrastructure is more complex than in the past, and if civil engineers are recognized only for their technical work, then the goals of the ASCE to promote "leadership by the civil engineering profession on public issues" and to enhance "recognition of civil engineering as a highly respected profession" will not be met and society will not be well served (ASCE 1999).

## Women and Minorities in Civil Engineering

Studies show that civil engineering will need more women and minority members in the future because of changing workforce demographics. Opportunities in the field are excellent. Because civil engineers have a close connection with the public, the profession should reflect the distribution of groups in the general population. Efforts to attract women and minorities to the profession must increase. Although educators and organizations are working on this issue, more efforts are needed. A study conducted by Colorado State University revealed some insight into the problem of attracting more women to the profession (Grigg et al 1997). Beginning in the late 1960s, efforts to attract more women into science, engineering, and mathematics increased because women were underrepresented; not only was the situation inequitable, the nation was losing a lot of good talent.

According to statistics published by the Engineering Workforce Commission of the American Association of Engineering Societies, 17.4% of women graduates in engineering in 1995 received bachelor's degrees; 16.7% master's and professional engineer's degrees; and 12.1% doctoral degrees. Unfortunately, these percentages have risen by less than 5% over the past 15 years (American Association of Engineering Societies 1996; Grigg et al 1997).

Although women constitute about 46% of the civilian labor force according to 1990 census data (National Science Foundation [NSF] 1995), only about 9% of engineering employees were women. Given that only 4% of total employment in the United States is made up of engineers, female engineers are a rarity in the work force (Grigg 2000). Although 17% of engineering graduates today are women (American Association of Engineering Societies 1995), it will be many years before these numbers are evident in the work force. The NSF Survey of College Graduates (NSF 1995) indicated that the highest representation of women is in chemical engineering (13%) and the smallest in electrical engineering (5%) and mechanical engineering (5%). Civil engineering (where most engineers in public works come from) is 8% female, although this number is skewed by the higher representation of women with a concentration in environmental engineering than those with concentrations in structural engineering, transportation, or city planning.

The facts seem clear. Women are underrepresented in engineering education, especially civil engineering, which is the usual path to public works; and there is agreement on what to do about it. All civil engineers, and especially educators, should adopt a more sensitive environment and promote networking among students; emphasize that engineering is a good background for other fields (e.g., law and public policy); provide mentoring and participation in professional societies; provide for personal contact among faculty, students, and potential students; have recruitment programs that include contact with parents and high school teachers or counselors; and provide female role models (Daniels 1988).

## A Final Word on Careers

Civil engineers receive plenty of career advice, ranging from which discipline or occupation to select to which courses to take, how to prepare a resume, and how to be successful on the job. Although all the advice is valid and useful, if any of it stands out it is to learn about civil engineering as a unique profession with many differences from other fields. The years ahead will be exciting ones, and civil engineers have much to contribute and can be rewarded well by preparing themselves well and acting professionally in the industries and situations in which they find themselves.

# 5 Engineering Design and the Infrastructure Life Cycle

## Introduction

The design process, which is the focus of civil engineering work, is a complex and multifaceted engineering activity. The work of civil engineers in design and construction sets them apart from other types of engineers. This chapter describes the design process.

When engineers hear the word *design*, they may think of a fixed, narrowly focused concept. When engineers think about the design of bridges, for example, they might think about selection of the structural form and materials of the bridge. Selection of form and materials is an element of the design process; however, the design process encompasses many additional elements.

Design is a tangible product, and the process that produced the product is the design process. The design of a bridge can be seen in a set of drawings. When the concept of design is expanded, all elements of the design of the bridge come into play, including the number of lanes the bridge should have, the projected mix and weight of the traffic, the reasons a particular location on the river was chosen as the bridge site, the reasons a bridge was chosen instead of a tunnel, and the reasons a highway was needed rather than a rail system. These aspects of design are complex, and it might be argued that they are planning questions rather than part of design. That argument creates a split between planning and design; however, design and planning are not separate activities. Instead, they are two points on the infrastructure life cycle continuum, which also includes construction, operation, rehabilitation, replacement, and demolition and decommissioning.

Much of the civil infrastructure in the United States has been built since World War II, and much of this infrastructure will soon require rehabilitation and replacement. It is not yet known how this rehabilitation can be done with minimal interruption of existing services. Replacing a bridge, for example, can become a major traffic disruption. To begin to understand, the design process must be viewed as encompassing the infrastructure life cycle—that is, an integrated, continuous process from the inception of the idea through the continued functioning of the constructed system.

Although design involves discrete tasks (such as drawing and measuring), the overall concept of design is broad. The philosophy of the design process covers topics from creative thinking (Petroski 1996) to specific procedures (US Bureau of Reclamation 1973). Literature on design ranges from social science and urban planning to how engineers prepare site plans. Design can even include rehabilitation and repair of infrastructure (Molof and Turkstra 1984). Neil S. Grigg compiled these various stages into a conceptual infrastructure management system (Grigg 1988). The ASCE has prepared a comprehensive manual on quality of the constructed project, including the design component (ASCE 1990). These themes must be pulled together to have a comprehensive view of the design process.

## Uniqueness of Civil Engineering Projects

Civil engineering projects are usually unique, in the public sector, and large in scale. The planning and construction of these projects have traditionally not considered maintenance, replacement, or decommissioning very well. The United States is now paying for past neglect through disruption of public services when highways, bridges, sewer systems, or other facilities are rehabilitated or replaced.

Viewing the infrastructure life cycle as a continuum (Fig. 5-1), specific points in the process can be identified. Engineers are involved throughout the infrastructure life cycle, including planning, conceptual design, preliminary design, final design, construction, operation, maintenance, and rehabilitation/replacement or decommissioning/demolition. All these stages revolve around the role of infrastructure in serving society.

## Planning

Planning is a flexible process and might precede the earliest phases of design in the form of thought, brainstorming, discussions, or rough

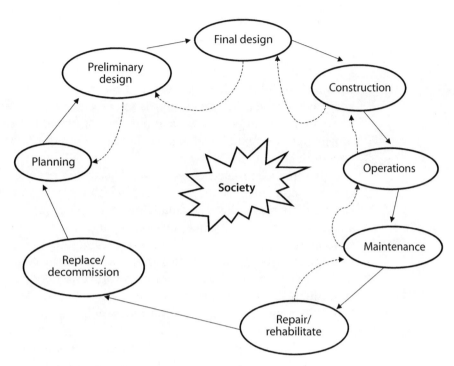

**Figure 5-1** Infrastructure life cycle.

sketches. There are many types of plans. For example, planning can refer to the activities of government agencies, such as maintaining master plans that form the framework for projects.

As the ASCE manual *Quality in the Constructed Project: A Guide for Owners, Designers, and Constructors* points out, the formal relationship between owners and design professionals creates a sharp delineation between the earliest phases of planning and the point at which the design professional begins a plan that becomes part of the construction process (ASCE 1990). Many discussions and exchanges of concepts take place in the realm of planning before the formal process begins, including those that occur in-house by the owner's team.

Conceptual design, usually part of the planning phase, is the point in the process at which the need for a project is identified and the first set of options is developed. This stage is important because it is when many of the project's options are identified and screening is done. Engineers are often criticized for focusing project options toward structural solutions (as opposed to nonstructural solutions), but it is important for engineers to be broad-minded in identifying all types of options, both structural and nonstructural. Flood plain work is a good example of this. Engineers do not always build a flood control dam; instead, they may zone the flood plain to keep people from living in dangerous areas.

Potential project options must be screened systematically, and the procedure used to eliminate options must be documented. To do this, the objectives of the project must be clearly identified and measurable criteria to compare the options formulated. Multi-criteria decision analysis tools are useful for this process. It is not necessary to have a detailed analysis for each criterion as long as the procedure uses the same level of accuracy for all options. For example, the estimated construction cost of project options should lead to a smaller set of options to be considered in greater detail.

Planning for projects covers several activities, each leading to convergence of concepts for the selected final concept. Planning studies and reports converge in level of detail and certainty of project characteristics and include reconnaissance, feasibility, and definite project plans. The content of these studies and reports varies with the application. The reconnaissance phase identifies projects that meet goals established in the overall planning and development process. It typically will lead to recommendations for further studies rather than to definite plans. The feasibility stage establishes definite feasibility, including financial, technologic, environmental, and political. This very substantive phase may result in documents that are costly to prepare, depending on the complexity of the project. The definite project phase results in plans, specifications, and operating agreements—all the guidance needed to construct and begin to operate the project.

The terminology for planning varies. Water projects will seem different from road projects because of differences in terminology. In reality, however, the processes have only minor procedural differences.

## Preliminary Design

In the preliminary design phase, a selected set of options is assessed in more detail. This requires a more detailed investigation of the options that remain under consideration. It may be tempting to use the conceptual design phase to select the preferred option and then jump directly to the final design phase, but the preliminary design phase offers field investigations that might be too costly for the number of options under consideration in the conceptual phase. The field investigations may reveal information that radically alters the feasibility of a specific option. An outcome of the preliminary design phase is a rating of the options from the standpoint of engineering, financial, economic, and environmental feasibility. Again, the use of a systematic methodology for analyzing options and selecting the preferred option is essential, including public involvement. The output of the preliminary design phase is the selection and scope of the preferred project option to be implemented.

## Final Design

The final design phase is used to develop the design details and drawings, as appropriate, for the selected option. This is the phase most commonly thought of as design. Although some of the calculations will have been completed in the previous phase, the bulk of the detailed analysis is done in this phase. The design process involves creative decision making about the configuration and details of projects, including the specification of materials, dimensions of project components, and details of member connections. There are many decisions to make that require experience and consultation. The results of design will be the detailed engineering documents and engineering drawings necessary to initiate the construction process.

It is important to acknowledge the risk inherent in the design of civil engineering projects and introduce concepts of risk analysis. Returning to the example of the bridge, the projected traffic and traffic loading are stochastic variables and the footings and bridge piers will be designed considering a design flood, another stochastic variable. Adding wind and earthquake loads yields two more stochastic variables. The design and analysis will normally use standard procedures and codes, and engineers should understand how these consider risk.

The engineering design will usually be a composite of work by various engineers or engineering design teams. One engineer or group of engineers might work on the footings for the bridge, another on the bridge substructure, and a third group on the bridge superstructure. The work of each of these groups must be integrated into the overall design, requiring a high degree of coordination and cooperation among the various teams to meet the project specifications on time and on budget. Figure 5-2 shows an engineer at the computer, where a large part of today's design process occurs.

## Design Review

After designs are completed, design review tools (e.g., value engineering) can be used to check whether the design is optimal for the cost. Tools have been identified and are used by various facility management organizations such as the Construction Industry Institute, the Business Roundtable, the Federal Facilities Council, and the National Research Council. One way to summarize these tools is by viewing best practices in design review. The Federal Facilities Council report lists tools and accompanying best practices (Table 5-1; Spillinger 1999).

Design reviews have the attraction that if applied early in the design of a project, substantial savings may be realized. For example, value engi-

**Figure 5-2**  Engineer performing computer design.

neering, one of several methods of cost reduction, analyzes functions and asks whether methods, processes, and materials that have been in use for years could be replaced by more economic elements. Value engineering seeks to answer these questions:

- What is it?
- What does it do?
- What must it do?
- What does it cost?
- What other material or method could do the same job?
- What would the substitute material/method cost?

## Construction

The goal of the construction phase is to build the project as close as possible to the final design, including cost and all other constraints. The close linkage of the final design and construction phase cannot be overemphasized. It is crucial that the designed project be constructable. If the project is not constructable, the design will have to be modified. Should this occur, information needs to be fed back into the design process so that future problems are eliminated. Even if the design is constructable, changes may occur during the construction phase. A common reason for changes is the unavailability of the specified construction materials. Another common reason is the discovery of an unanticipated field condi-

**Table 5-1**  **Design review best practices.**

| Issue | Best Practices |
|---|---|
| Owner's role | — Be a smart buyer with an adequate in-house staff.<br>— Develop a scope of work that fully defines the owner's needs.<br>— Avoid the temptation to micromanage design reviews. |
| Teamwork and collaboration | — Use teambuilding and partnering to build good working relationships.<br>— Involve all interested parties in design reviews from the inception of planning and design.<br>— Use the same A/E firm throughout the facility acquisition process to maximize continuity.<br>— Use senior, experienced personnel to evaluate the design process and guide the review.<br>— Avoid changing participants during the review process.<br>— Participate in a design awards process to recognize and reward excellence. |
| Advance planning | — Focus attention on review early in the process to maximize improvement.<br>— Do not start plans and specifications until preliminary engineering is complete. |
| Process | — Tailor design review to project specifics.<br>— Maintain momentum and keep facility acquisition on schedule.<br>— Attend to interfaces between civil, structural, electrical, and mechanical facets.<br>— Exploit technology, especially information technology.<br>— Conduct a postoccupancy review to identify lessons learned. |
| Benchmarking | — Assess benefits of design review.<br>— Document unusually good and bad performance. |

tion (e.g., a geologic condition that requires a change in the design of footings for the bridge piers). All construction changes must be evaluated to ensure that the design integrity of the project is not compromised. Figure 5-3 illustrates construction under way on an interstate highway project.

The initiation of construction involves numerous legal and procedural steps. These activities are managed by an engineering staff of a public works organization or other owner. The staff undertakes a number of related functions, ranging from initial surveys and plans to construction management and record keeping. The construction process involves bidding, review, award, organization, construction itself, inspection, and

**Figure 5-3**  Interstate highway construction. (Courtesy Colorado Department of Transportation, 1999; photo by Gregg Gargan)

acceptance. A formal process for these steps has been developed over many years as a requirement to control costs, quality of construction, and quality of the final product. The construction phase begins with preparation of the contract documents. The actual construction involves complex operations that must be designed to fit the infrastructure situation involved. An important step in the construction process is the quality control (QC) and quality assurance (QA) process, which culminates with the final inspection and acceptance activity.

## Project Management

Construction project management is an essential element in the solution of infrastructure problems. Project management as a generic skill extends to many other types of projects as well (see Chapter 6).

Unless construction projects are completed with quality results and within budget, there can be no solutions to the problems of infrastructure. The actual management of multifaceted projects is a complex undertaking and there is no direct, easy recipe for success. An example of a project that is even more complex than most construction tasks is the Manhattan Project (which resulted in the first nuclear weapon) or the Apollo space project (which put the first man on the moon).

Project management as a skill involves many complex activities. A few sample definitions can help to interpret it (Colorado State University 1984).

- A project management system (PMS) is a networking system that uses an integrated approach to successfully control and direct a project.

- A project is a definable concept that employs the functions of planning, design, finance, construction, and control to achieve an end.
- Management is the catalyst that directs and guides the functional operations to complete the project on time, within budget, and at an acceptable quality level.
- A system is an information network that integrates all functional requirements through various management levels for the successful direction and guidance of the project to an efficient end.

Project management tools that can improve overall effectiveness include the following:

- A good definition of project scope
- Clear roles and responsibilities
- Effective documents for design and finance procedures
- Valid estimates
- Activity-based schedules
- Cost-control systems
- QA/QC systems
- Information handling systems
- Change order handling systems

The role of the infrastructure manager and the engineer in inspection of constructed work is obviously important in determining the final quality of work. Inspection determines that the work has been completed according to the plans and specifications. The ASCE has begun to emphasize quality in construction as part of its leadership role in the construction industry (ASCE 1990).

## Operations

The operations phase is not well understood by some civil engineers because not all engineering structures require operation. For example, a section of roadway does not have to be operated, although a roadway system (e.g., a traffic signal system) may have to be. Many civil infrastructure projects are operated, including dams, locks, and water and wastewater treatment plants.

Projects may have engineering staffs to operate them. Engineers may be required to make operational decisions about complex systems, often during emergencies such as floods. Ultimately, successful operation is proof of the design. If the project cannot meet its objectives, then the design, construction, or both are flawed. As was noted earlier in this chapter, it is essential that operational difficulties are conveyed to the project design and construction engineers. Even smaller or less complex projects may require operational procedures that are developed by engineers. The design characteristics of a project will affect the operational guidelines.

Operations management is a discipline within industrial engineering that includes production management, facilities management, maintenance management, and information management. Classifications such as these, as well as specific studies of operations as a science, are available (Greene 1984).

Quality control has been emphasized by industry in recent years. It is better to do things right the first time than to have to correct mistakes. QC ensures that the quality of the product is within acceptable limits, as defined for that particular product. Sometimes QA measures are used in place of QC. Production operations offer ideas about QC. In his book *Operations Management: Productivity and Profit,* James H. Greene devotes most of two chapters to the subject, with one covering the statistical aspects of QC alone. His discussion of the internal organization for QC offers a test as to whether organizations are competent in making quality checks. The parts of the organization Greene identifies are inspection, administration and record keeping, QC engineering, and the gauge room (Greene 1984).

Quality control functions are particularly relevant to water or wastewater treatment functions because they are necessary to meet regulatory requirements. The principles can be applied even more vigorously than the regulations require, however. For example, if the goal of a water utility is to deliver high-quality water with certain pressure levels during certain times, it would need an inspection program, a QC sampling effort, laboratory work, and record keeping. Another example is storm water. If the goal is to prevent a certain level of flooding, how is success determined? The normal indicator might be the absence of complaints, but the organization could plan and execute an inspection program. QC programs can be considered as taking the offensive rather than being on the defensive.

## Maintenance

The maintenance phase of a project requires engineering expertise similar to that needed for the operations phase. Figure 5-4 illustrates a utility line being dug up and demonstrates the level of effort that may have to be devoted to maintenance, especially in congested areas.

Maintenance protocols and procedures are usually developed by engineers familiar with both project design and operational requirements. Deferred or inadequate maintenance of civil infrastructure is a major problem throughout the world. Another important function of this phase is to ensure that the project can be maintained as needed. For example, consider the maintenance of a storm drainage system. Storm

**Figure 5-4** Pipeline work. (Source: Colorado Department of Transportation, 1999)

drains, manholes, and other access facilities must be designed to allow maintenance on them. Not only is it essential to have adequate access for personnel and machinery for maintenance, it is also necessary to minimize the impact of maintenance on the operation of the facility. Closing a water treatment plant or a water main or even a bridge for an extended period for maintenance may be impossible. The planning and design of the project may require redundant operational capabilities.

Maintenance must be supplied at different levels, and this has implications for both the operating and capital budgets. Preventive maintenance heads off problems and is the most common maintenance operation in organizations. Corrective maintenance has minor and major aspects, depending on the extent of the correction needed. Corrective maintenance involves repair, replacement, and rehabilitation of facilities (sometimes called the three Rs of infrastructure).

Maintenance-related functions are condition assessment, inventory, preventive maintenance, and corrective maintenance. Preventive and corrective maintenance are sometimes called major and corrective maintenance, respectively, to stress the importance of some of the corrective actions. The condition assessment activity is a link between operation and maintenance functions and illustrates why the two functions must be unified. If the condition starts to worsen, the operation will be affected and it will be time to schedule repairs and maintenance.

The result of the inventory and condition assessment functions is determination of the location and condition of the system components. There should always be an answer to the question, "Is the system OK?" If it is, the operations and surveillance continue. If it is not, corrective and major maintenance activity must begin. The need for an inventory seems obvious for valuable equipment; however, the need for inventories of infrastructure systems must also be considered.

Corrective maintenance requires a decision: Is the deficiency serious enough to warrant entering the planning, programming, and budgeting activity, leading to a request for capital? If the problem is major, the budget activity incorporates information about new standards and growth forecasts to lead to decisions about rehabilitation or replacement. The planning part of the budget process also yields information to be used in the needs assessment process, which is the link between the major maintenance program and the capital budgeting program. Because of this linkage, it seems appropriate to have the same planning staff do the planning for rehabilitation work and new facilities.

The sequence of activities that leads to a decision about minor or major corrective actions also explains the differences among repair, rehabilitation, and replacement. These activities are part of a spectrum that covers the range from routine operating budget activities to major capital budget activities. This is an important distinction for management because the operating budget should have enough funds for routine and important repairs, and there should be sufficient funds in the capital budget for major rehabilitation without excessive deferral. The dividing point is somewhere between repair and rehabilitation, but each manager and organization must decide this individually.

Maintenance management systems (MMSs) have arisen in recent years in an attempt to bring together the disparate concepts of maintenance activities into a holistic approach to caring for a system. This is the systems approach applied to maintenance. An MMS is a program ensuring that overall maintenance is managed adequately. It involves all the tasks of management—planning, organizing, and controlling—and it requires an effective decision support system. MMSs include condition assessment, preventive maintenance, and corrective and major maintenance; and the decision support system will provide the information and data needed for these activities.

## Rehabilitation and Replacement

The rehabilitation and replacement process is one of the most challenging and exciting for today's engineering students. Much of the infrastructure developed over the past 50 years is in need of rehabilitation or replacement. The major concern in rehabilitating or replacing the current infrastructure is that it is in use and providing a service that must continue. Consider the previous example of the water treatment plant. If that plant is the sole facility treating water for a particular community, it is simply not possible to shut it down for an extended period for rehabilitation. Or consider the case of Hoover Dam. How will it be replaced when it provides water for portions of the western United States and Mexico?

The issue is similar to that of automobiles. If cars could be made so that they were never thrown away but simply rehabilitated, then much of the waste and energy requirements surrounding that industry could be eliminated. Not surprisingly, some are calling for such an approach today in order to reach a sustainable society.

It is the same with infrastructure, much of which was not designed with either rehabilitation or replacement in mind. As engineers are thrust into the rehabilitation and replacement process on highways, bridges, water and waste treatment plants, and dams, they will identify design options that could make the final phase of eventual replacement easier. This information should be fed back to the engineers working on planning, design, and construction. Not considering the problems of service interruption, there are still significant engineering challenges to be addressed. One example is the expected impact of a specific rehabilitation treatment on the project's usability and lifetime. Questions to answer include the following:

- What are the critical factors that determine whether a project should continue to be rehabilitated or whether it should be replaced?
- What are the critical factors that determine which of several potential rehabilitation strategies might be the most effective?

## Decommissioning and Demolition

The decommissioning and demolition process is the final phase of the infrastructure life cycle. Similar to the rehabilitation and replacement phase, this phase has not traditionally been considered in the design process. The growing awareness of life cycle costs and environmental consequences of infrastructure projects has brought this issue to the forefront. Consider the engineering challenges in decommissioning a nuclear power plant or a hazardous waste processing facility. For example, consider the demolition of a dam. The requirements for decommissioning and demolition need to be considered in the earlier phases because they may significantly impact the design and operation of the project. When sustainability is considered, decommissioning of old projects offers opportunities to do things better than in the past.

# Organization and Oversight of Design Activities

Design is done by a special engineering staff, who may be in-house engineers or outside consultants. Once a project is completed, it is turned over to an operating staff that usually works for the owner. Both public and private infrastructure organizations usually include an engineering department as a staff function. These might be the engineering department of an industrial firm, the city engineer's office in a city, or the base engineer's office in the military.

*Urban Public Works Administration*, published by the International City Management Association (ICMA) about public works management wrote that the engineering function (which would take place in all of these offices) includes surveys; studies and investigations; capital improvement program development; field survey work; planning, design, and cost estimation for construction; construction contracting and contract administration; construction inspection and supervision; preparation of maps, records, construction records, and reports; critical path method and program evaluation and review techniques (PERT) charts; and assistance in maintenance, repair, and reconstruction work (Korbitz 1976). The successor volume to *Urban Public Works Administration* listed processes that fall into the category of engineering and contract management: the planning and design process; construction management, including inspection; surveys, maps, and records; engineering management and private activity; and retention and use of consultants. Private activity in this context means interfacing with developers, consultants, and other private sector participants in the plan–design–build process (Martin 1986).

Infrastructure management includes three principal aspects of the engineering function: keeping the standards for infrastructure development, maintaining the quality of construction, and keeping records. Keeping standards sets the stage for quality development of facilities and is the greatest determinant of cost of infrastructure. The city engineer is responsible for both quality and cost. Political pressure from the private sector usually helps keep costs down. Maintaining the quality of construction is critical to ensuring that investments in infrastructure pay off. The public could not trust a public works official who accepts shoddy construction work or engages in corrupt practices, yet these problems occur in infrastructure. The final task, keeping records, is important to the development of an effective decision support system. The geo-database, where common mapping is used for all public utilities, would be the responsibility of the engineering office. All the as-built drawings, official standards, surveys, reports, plans, and related data would be located in this office. Figure 5-5 shows a completed portion of I-70 through Glenwood Canyon, Colorado. This project included a great deal of design review to deal with engineering and environmental questions.

## Computing and Analysis Tools

Computing and analysis tools are used throughout the design process, and their use is increasing. These tools fall into three general categories: data management and visualization, development of information through analysis and modeling of data, and presentation of information to facilitate decision making. The second category, analysis or modeling, can be divided between physical and mathematic analysis or modeling; and mathematic analysis or modeling can be further subdivided into simulation and optimization approaches.

### Use of Computing Tools

Each phase of the design process uses various combinations of computing tools. Databases and data visualization tools can be used in any phase

**Figure 5-5** I-70 through Glenwood Canyon. (Courtesy Colorado Department of Transportation, 1999)

and are particularly useful in the conceptual design, operations, and maintenance phases. For example, in the maintenance phase, MMSs develop records of system status and maintenance and predict appropriate maintenance schedules. Simulation models find application in all phases, whereas optimization models tend to be used mostly in the final design, construction, and operations phases. Tools for effective presentation are used throughout the design process.

## *Advances in Computing*

Over the past two decades, tremendous advances have been made in computing. There is a wealth of user-friendly, well-documented models for various aspects of civil engineering. General purpose software such as spreadsheets and databases make it fairly easy to perform engineering calculations and develop engineering models. Regardless of whether a model has been developed by the engineer or someone else, it is the responsibility of the engineer to use the model appropriately. Engineers are held accountable for the safety and functionality of their final product and are expected to ensure that all the resources they use, including models, provide reasonable and accurate results. Engineers are then responsible for making sure that any models used provide correct results.

## *Example*

The example of the highway bridge can assist in understanding what computing tools might be needed at the conceptual design stage. Basic topographic information could be obtained from maps of the site as well as general information on the type of soil in the area. These data would likely be stored in a database. Because these data are convenient for display as maps, GIS software could be used to display the data. Databases of existing traffic in the surrounding area would also likely be used. The database and GIS software are examples of tools in data management and visualization.

## *Models*

Mathematic models are very useful in engineering analysis and can be generally defined as mathematic representations of the behavior of physical systems. The broad definition would include either a single equation or a set of equations that are used to represent a physical process. Inputs are provided and the model is solved to predict outputs.

Most models are simulation models, meaning that the user manually changes the inputs to produce the desired outputs. An example would be a structural model in which the user inputs the loads and desired material and the model computes the specifications of the structural members.

The user can adjust the structural material until he or she is satisfied with the structural member specifications. Another example would be a hydraulic model in which the user inputs the channel shape and dimensions and the model computes the flow velocities. The user can change the shape and size of the channel until the predicted flow velocities are in a desired range. An optimization model is one that automatically changes the inputs to produce an output that is best, according to a specified mathematic goal. A common example of an optimization model is least-squares regression analysis. In this case, the user asks the model to find the best values for the parameters that minimize the sum of the squared errors of the model prediction. An optimization model can be thought of as a simulation model with an automatic trial-and-error search method.

## *Statistical Tools*

Statistical tools could be used to estimate the future percentage of heavy vehicles traveling on a road or highway based on analysis of existing traffic data. These estimates could then be used in a simulation model that computes structural loading due to traffic. The predicted structural loading in addition to the desired construction material could then be used as an input into another simulation model that would compute the specification of the structural members. A structural optimization model might be used to find the specification of structural members that minimizes the cost or weight of the structure. These models are examples of the development of information.

## *After Models*

After the models have been used, the results would need to be synthesized to facilitate comparison of the various options, such as alternate construction material or alternate bridge locations. Either databases or spreadsheets might be used to organize the information into tables or graphs. This would be an example of the effective presentation of information. If a large number of options are to be evaluated, a multicriterion decision analysis (MCDA) model might be used assist the decision makers in identifying the most desired options. The MCDA procedure is a systematic scoring process, similar in concept to the scoring procedure used in a university course. It can be thought of as a simulation model in which the inputs are the relevant decision criteria and their relative importance determines the outcomes of the decision analysis and recommendations.

## *Errors in Models*

Potential errors in a model can be classified as formulation, implementation, or application errors. A formulation error refers to a mistake in the basic equations the model will solve. These equations could have errors in their underlying theory, they could contain typographic errors, or they may be incomplete. An implementation error has to do with the coding of the model into some solution software or computer language. An application error refers to applying a model to a problem for which it was not intended or extending the use of the model outside of a valid range.

For example, suppose an engineer develops a model for sizing a culvert to be placed under a roadway. Further, suppose the engineer codes this model in a spreadsheet and uses an equation from a hydraulics textbook that contains a typographic error in a coefficient of the flow equation. Even if the model is coded perfectly and applied to a valid range of data, the results will be in error. Conversely, if the equation in the textbook is correct and the engineer makes a coding mistake that results in a calculation using an incorrect coefficient, the results could be similarly flawed. The formulation and coding may be completely correct. If the engineer applies the model to a condition for which it was not designed (e.g., unsteady instead of steady flow), the result will again be in error.

The processes engineers use to help eliminate or at least minimize errors in models are verification, calibration, and validation. Verification implies testing a model to help ensure it is computing correctly. Verification aims at identifying formulation and implementation errors. There are various methods that can be used for verification, including the common approach of using the model to solve one or more standard problems with known answers. These standard problems can be called *benchmarks*. These benchmarks might be solved by hand by the engineer, be available in textbooks, or be solutions from other similar models. Another approach is to test the model at its limits. For example, if the inputs of a model were set to a limit (such as zero), the model would be expected to respond in a certain way. In the example of the culvert design model, a zero flow should result in no culvert being required or a culvert with zero size.

Calibration is the process adjusting any appropriate coefficients in the model, based on observed data, to describe the specific situation being analyzed. As defined, a model is a conceptualization of a physical system or process. Often, all the complexities in a physical system use simplified rate equations to describe the processes that are not understood. In the example of the culvert design problem, a coefficient could be used to describe flow resistance in a culvert due to the material that comprises the culvert. To construct a concrete culvert, the appropriate flow resistance coefficient must be used. Further, if a specific concrete culvert is modeled with observed data, the general coefficient for concrete culverts might need further adjustment to more closely represent the characteristics of the specific culvert in question. The basic philosophy in the calibration process is to use the observed data as model input and adjust appropriate coefficients until the model results match the observed results as closely as possible. This calibration process not only helps engineers ensure that the model coefficients are correct but can also help identify formulation, implementation, and application errors. If a model cannot be calibrated, or if the calibrated parameters are outside an acceptable range of values, this indicates a flaw in the model. This flaw could be the result of errors in either the model's formulation, implementation or application, or a combination of these errors.

Validation is the process of ensuring the model is appropriate for solving the kinds of problems it is intended to solve. R.M. O'Keefe succinctly delineated between verification and validation by stating that "Verification is building the system right. Validation is building the right system" (O'Keefe et al 1987). Validation is primarily focused on application errors, although it can be useful to identify formulation and implementation errors as well. A commonly used process in validation is the split sample approach. For example, a set of observed data might be divided in half. The first half of the data would be used to calibrate the parameters of the model, which would then be applied to the second half of the observed data to see how it performs. Ideally, the coefficients of a model should be essentially constant over space and time and the model should perform well over the second half of the data. If the split sample approach shows that the model does not perform well, this indicates a flaw in the model.

The validation process is also useful in determining whether the output from the model is sufficient for answering the questions the model is intended to answer. Any model is only a tool to assist engineers in making a decision. The model provides information the engineer can incorporate into the decision process. The model must then provide the required information in the required format for it to be useful. Often a major consideration in selecting or designing a model is the ability of the model to provide information in a readily understood format.

## *Summary of Models*

It is the responsibility of the engineer to exercise reasonable care in the development and use of models. Engineering models are an integral tool used throughout the design process. Models should not be presumed correct unless they have been thoroughly tested, and all models should be tested in some fashion before their results are used as part of the design process.

## *Use of Computers for the Presentation of Information*

Today, the use of computers to present information is expanding rapidly. With advancing presentation techniques, greater Internet capacity, and the development of new visualization techniques it is hard to predict how presentations will change in the future. Some see the impact on education as so great that they have coined the term *edutainment*, meaning the convergence of education and entertainment techniques. Regardless of these advances, engineers will need to ensure that presentations are based on solid methods of engineering investigations, analysis, synthesis, and decision making. Then, a variety of computer-based techniques can be marshaled to prepare the presentations.

## Design Codes and Standard Design Procedures

Design codes and standard design procedures are widely used in engineering. It is not necessary or desirable to develop an analysis for each project. The experience of the profession can be used to document the accepted solutions to standard problems. These accepted solutions have been developed using well-tested models and have been validated in the field. They usually consider risk and uncertainty through the use of safety factors. These accepted solutions are then made available via design codes. The structural design codes are a good example of this. Civil engineers compute the loading for a building and then use the structural design codes to identify the appropriate size of the structural members.

Design codes often become legal constraints (e.g., by incorporation into building codes for urban areas). These building codes represent the standard for design requirements. A building design should meet the building code and can exceed the code, and the design should not violate the code except under very unusual circumstances.

Standard procedures might apply to design, maintenance, repairs, or operations. These represent the state of the practice in terms of the steps and the order of those steps that should be followed to help ensure a good engineering product. The design of a highway culvert is a good example. State departments of transportation develop design manuals for highways that include suggested procedures to determine the size and shape of highway culverts. This procedure will list the steps in the process, including recommendations on the use of specific models and design codes.

The engineer should consider risk and uncertainty during the design process. As noted, design codes usually embody factors of safety to account for some of the stochastic aspects of the project. This risk can be defined as the probability that an engineering project will not meet its demands over a specified time period, and a factor of safety can be defined as the ratio of the capacity to the demand for a project (Kottegoda and Rosso 1997).

In the example of the bridge, an engineer might calculate the expected load for the bridge yet design for a higher value, thereby providing a margin of safety. That calculation, however, might be based on a current load limit for trucks. If the load limit for trucks were to significantly increase during the life of the bridge, then the true margin of safety might be less or the load capacity of the bridge exceeded.

Defining failure is not a simple task because of the various consequences of failure and the impacts of those consequences. One definition is that failure is the inability of a project to meet its requirements (Kottegoda and Rosso 1997). For example, a bridge might fail to convey a large volume of traffic because its capacity to support the traffic flow is insufficient. The consequences of such a failure will be delay and inconvenience to the travelers. Over time the traffic flow will decrease, and the bridge will return to a satisfactory state in terms of traffic capacity. If the bridge fails structurally and collapses, the consequences could be serious, including loss of vehicles and injury or death to their occupants.

During the design process, engineers attempt to identify possible sources of failure and to quantify the probability associated with those sources. Designers want to provide the capacity in the project to meet the project requirements within a specified probability of failure. For example, a designer of a storm water detention pond may design the pond to be overtopped only if the runoff exceeds that of a 50-year storm. They may further design the pond in such a way that even when it is overtopped it can safely convey the excess flow without collapsing the dikes of the pond up to the 100-year storm. Simulation models are used to evaluate the consequences of failure. These models can be run with many input data sets and the results statistically analyzed to develop exceedance and non-exceedance probabilities of various kinds of failures. Modern computing capabilities are enhancing the ability of engineers to more effectively consider risk analysis in the design process.

## Design by Consulting Engineers

Engineering normally costs less than 1% of the life cycle cost of projects; therefore, it pays to get good engineers. In public sector work some engineering design is done in-house, although a great deal is contracted out, especially capital facilities design. One of the tasks of the engineering function, when it represents the owner, is to retain and supervise consulting engineers.

There are advantages to doing work in-house and to using consultants. With in-house work, there is better control over details and sometimes reduced costs. With consultants there is a wide range of engineers, and they can be used only when needed. Sometimes consultants can be more cost effective than in-house staff. This has to be evaluated on a case-by-case basis.

There are approximately 10,000 to 20,000 consulting engineering firms in the United States, depending on how they are defined. These firms are organized into state consulting engineering councils and nationally through the ACEC, which is headquartered in Washington, DC. The business category of consultants includes more professionals than consulting engineers—management consultants, financial consultants, independent practitioners, lawyers, and others who offer related business services. Consulting engineers are the main group involved in constructing the nation's infrastructure; this group is largely responsible for its planning, design, and supervision. *Consulting Engineering: A Guide for the Engagement of Engineering Services* presents guidance about the consulting engineering profession (ASCE 1996):

- The practice of consulting engineering (professional responsibility, client–engineer relationships, selection of a consulting engineer)
- Classification of engineering services (feasibility investigations, appraisals and valuations, preliminary design, operation)
- Guidance on the selection of an engineer and available selection procedures; methods of charging for engineering services (hourly billing rate or fixed price)
- Total project cost (legal and administrative costs, contingency allowance) and contracts for engineering services (contracts with associate professionals, limitation of risk, and partnering)

Other services mentioned in the predecessor to *Consulting Engineering* include the following (American Society of Civil Engineers, 1996):

- Field investigations
- Data collection
- Environmental impact assessment
- Report preparation
- Impact statements
- Design services
- Specification preparation
- Bids
- Observing construction
- Testing and evaluations
- Appraisals

The use of consultants involves selection, contracting, management, and compensation, which are responsibilities of the owner (normally an infrastructure management organization). Selection of consultants is usually through competitive negotiation. This procedure, advocated by ACEC and in compliance with the restraint of trade requirements of the federal government, provides an alternative to the lowest-bid process, which is not seen to lead to the highest-quality work or the best arrangement for either client or engineer. The process of competitive negotiation has superceded the earlier practice of simply selecting firms without a structured decision process.

Competitive negotiation involves the following process. The client asks engineering firms to submit qualifications and performance records for evaluation. Factors to be considered in reducing this lengthy list to a short list are technical qualifications, experience in similar projects, reputation, timeliness, mobility and workload, and financial references. All firms that make the short list are asked to provide brief presentations explaining their concepts of the work to be done. This allows the firms to express their creativity in meeting the client's needs. On the basis of these presentations the client ranks the firms and begins a negotiation with the top one. The negotiation involves the scope of work, other contract provisions, and compensation. If the negotiations are successful, the contract is drawn. If not, the next firm on the list can be asked to begin negotiations.

The request for proposals (RFP) used in the selection of consultants is a work of art. The client uses the RFP development process to clarify what is wanted, and the RFP provides a clear statement to the engineer and all others involved as to the scope and objectives of the project. The preparation of proposals for the engineering work is also a work of art. The engineer uses the proposal development process as a way to express creative ideas about the solution to the client's problem as well as a way to present credentials.

The management of consultant activities begins with preparation of the contract that specifies the scope of work and all items that are required for delivery. Also, regular reports are required from the consultant. Frequent meetings help to maintain coordination. The consultant should be viewed as an extension of the client's staff in one sense and as an independent contractor needing direction in another sense.

## Quality in the Constructed Project

The infrastructure life cycle is deeply affected by quality in the constructed project. This has been recognized by the ASCE, which has provided a manual on quality in the constructed project.

After the failure of two walkways in the Kansas City (Missouri) Regency Hotel in 1981 and in recognition of a number of other less-visible failures, a Structures Failure Conference was held in 1983. In November 1984, a group of nearly 100 members of the design and construction industry met in Chicago to discuss how to reduce design flaws, accidents, cost overruns, and similar problems. Participants were concerned about rising liability costs, standards of performance by which construction professionals are judged, and relationships among members of the construction team. As a result, it was agreed that the ASCE would prepare a comprehensive manual on quality in the constructed project.

The ASCE organized a steering committee, composed of 40 authors and 90 reviewers. Committee members are listed in *Quality in the Con-*

structed Project: A Guide for Owners, Designers and Constructors (ASCE 1990). *Quality in the Constructed Project* is an impressive roster of experienced construction industry professionals. It also lists recommendations on principles and procedures that lead to quality in the constructed project and guidance. Recommendations include the following (ASCE 1990):

- Roles
- Responsibilities
- Relationships and limits of authority for project participants
- Levels of performance
- Acceptable standards of quality
- Principles of communication
- Procedures for design and construction
- Management practices
- Teamwork
- Expectations and objectives
- Value of mediation
- Stressing of project benefits over individual team member benefits
- Owner's selection decisions
- Peer review
- Shop drawings
- Standard forms and practices

*Quality in the Constructed Project* also provides guidance for the principal members of the project team: owner, design professional, and constructor. A traditional relationship among these parties is one in which the owner conceives a project and selects a design professional to prepare plans, leading to bidding and the construction process. Today, new forms and relationships for project organization are being developed, such as turn-key projects, where one firm has responsibility for all phases, construction management, and owner construction. With today's changing business climate and a focus on privatization, build–operate–transfer (BOT) schemes are becoming more prevalent.

Regardless of their relationships, each member of the project team must accept responsibility; strive for economy and efficiency; cooperate and coordinate; and adhere to budget, schedule, program, and high-quality standards. Selection of the design professional is a key ingredient to success in the project. The ASCE recommends a process whereby the design professional submits statements of interest and qualifications on the basis of the owner's requirements and invitation. After the owner selects a design professional on the basis of qualifications, negotiations (including compensation) can proceed. The ASCE believes that the best agreement results from establishing a fee after extensive scoping discussions occur. Legislation at the federal level (Brooks Law) and state levels has reinforced this procedure.

Quality in the constructed project is a term used to integrate the many aspects of the construction process that lead to success in meeting the owner's requirements. *Quality* is a very broad term. For example, if the owner is the public, then the owner's requirements may extend to quality performance through a life cycle that lasts many years. The ASCE lists the characterizations of quality as meeting the owner's requirements for functional adequacy, completion on time and within budget, life-cycle costs, and operation and maintenance.

To visualize this, consider that someone has contracted to have a family home built. This person will be partially satisfied if the home achieves all functional requirements (e.g., enough space, the right rooms). The person will be more satisfied if the construction team delivers the home on schedule and within budget; if all costs (including maintenance and final sale price) are on target; and if the building works in operations and maintenance (e.g., utilities, repairs). Such concepts are complex enough for a family home; imagine how complex they are for a mega-project.

Quality in design and construction means that the design professional is given the means and guidance to do a good job, including a clear scope of the work, enough budget, timely decisions by the owner, and a contract to work at a fair fee. The constructor requires clear plans and specifications, enough detail for the firm to prepare an effective bid, and timely and fair decisions by the design professional on matters related to contract administration. The other main consideration is compliance with the regulatory requirements relating to public safety and health, environmental considerations, protection of public property and utilities, and conformance with all applicable laws and regulations.

As the leader of the project team, the owner has two main responsibilities: to develop workable objectives and to communicate clearly to the others the role and responsibilities of the owner. Effective communication is an important ingredient in the success of constructed projects. Information such as requirements, expectations, scope, costs, schedules, and technical data is vital to team members. Coordination is also vital, and success in coordination, based on effective communication, is the key. According to the ASCE (1990), insurance studies show that most legal actions taken by owners are not based on aspects of the final project but rather are communication issues, such as surprises, frustration over unaddressed problems, lack of positive personal relationships, or not being informed about problems.

According to the ASCE, the design process should include such good practices as office operation, designer–owner–constructor relationships, quality programs, good design procedures, and compliance with regulatory requirements. Peer review can add value to the design process and, according to the ASCE, is the "highest level of action to improve quality in design of constructed projects" (ASCE 1990).

Some construction documents are described in various publications of the Engineer's Joint Contract Documents Committee. The ASCE has published a number of other forms and documents as well, including these listed in their online publications catalog (ASCE 1999):

- Agreement among parties (owners, designers, and project peer reviewers)
- Joint venture agreement
- Standard general conditions of the construction contract
- Application for payment
- Bid bond
- Certificate of substantial completion
- Change order
- Construction payment bond
- Construction performance bond
- Engineer's letter to owner requesting instructions concerning bonds and insurance for construction
- Notice of award
- Notice to proceed
- Owner's instructions regarding bidding procedures and construction contract documents
- Standard form of agreement between engineer and geotechnical engineer for professional services
- Standard form of agreement between owner and contractor on the basis of cost-plus forms of contracts
- Standard form of agreement between owner and contractor on the basis of a stipulated price
- Standard form of agreement between owner and engineer for study and report professional services
- Standard form of agreement between owner and design/builder on the basis of cost-plus contracts
- Coordinated multiprime design agreement between owner and design professional for construction projects

According to *Quality in the Constructed Project* (ASCE 1990), responsibilities during the construction process begin with the owner, whose responsibilities (except in emergencies) include activity coordination, contract enforcement, and stopping work. The design professional is responsible for design changes and interpretation of construction documents. The constructor is responsible for construction methods, direction of labor, and job safety.

# 6 Management for Civil Engineers

## Introduction

Civil engineers work more in management than do other types of engineers and can assume a variety of management roles, including project manager, public works manager, and consulting firm manager. After working in some of these roles, civil engineering graduates often advise educators to teach more business and management skills. This chapter outlines the field of management within the context of civil engineering practice. Although there is vast literature on this subject, much must be learned from experience.

An NSF survey showed that most civil engineers identify management as their top activity, followed next by design work and then computer applications (Burton et al 1998). This statistic introduces a dilemma: The primary concern of civil engineering education is design, but their primary activity is management. It is important to remember, however, that it is primarily engineering activities that are being managed.

Civil engineers manage many things, including people, resources, activities, and teams in both private firms and public agencies. In addition, they manage construction projects and business contracts, and they develop management systems (e.g., a floodplain management system in a city). They also manage public sector organizations (e.g., city or public works agencies), private sector firms (e.g., consulting firms, contracting companies), and facility operations (e.g., traffic controls, water treatment plants). Further, civil engineers apply management skills and techniques to a variety of other situations, including working with environmental regulatory programs and volunteer organizations, problem-solving and consulting situations, and financial management.

# Defining Management

The term *manage* means to direct or control the use of something using authority, discipline, or persuasion (*Webster's II* 1984). Management is the act of managing, supervising, or controlling. Management is distinctly different than design, analysis, reporting, and other tasks that civil engineers are trained to do and requires a different skill set. Management involves decision-making activities surrounding goal selection and attainment.

In a practical sense, managers allocate resources to achieve objectives; and they use a collection of techniques to plan, organize, and control individual and group activities. They require skills and knowledge that involve the areas embraced by business administration, public administration, and related fields (e.g., management theory and cases, organizational theory, finance, public relations and marketing, communications, and computer information systems). Some of these management topics are covered in other chapters in this book, and in courses in the civil engineer curriculum. Most civil engineers who become managers, however, do not have formal training in it—they learn by doing.

Management activities of civil engineers often mean different things because the work varies so much. For example, civil engineers can assume such roles as project manager, business owner, regulator, professor and academic, public works manager, federal agency manager, sales, and the military. Figure 6-1 illustrates a high level of management: a meeting of an interstate river commission chaired by a state governor.

Consulting engineers might work as project managers, requiring knowledge and skills in project management. Consulting engineers might be called on later to be involved in the company's management activities (e.g., setting up a company, working with a board of directors, issuing stock, supervising employees). This requires skill in business management. If the company sets out to develop a new strategy, the management task might be strategic planning. Civil engineers working in local government will be confronted with another group of management activities, including developing contract documents, setting up PMSs, searching for new employees while following public hiring rules, and directing an agency or division of an agency. This requires skill in public sector management (or public administration). For engineers working in a utility or a transportation system, for example, the management task might be to run a traffic system or treatment plant, requiring operations and maintenance management. In the federal government, civil engineers might be

**Figure 6-1**  Meeting of the Delaware River Basin Commission. (Courtesy Delaware River Basin Commission, 1994)

involved in public administration at a different level or in policy work with congressional committees. Civil engineers might be involved in budget work, liaison with the Office of Management and Budget, or setting up and staffing a regional office. Other management activities could include volunteer work with an association, perhaps serving as a committee chair or director. Management also occurs in personal settings, such as organizing work and accomplishing personal goals.

This chapter introduces topics to assist in gaining management skills, whether from a formal education or learning by doing. To get a good product, you need process, organization, leadership, and bottom-up commitment (Creech 1994). Keeping this concept in mind can assist civil engineers in focusing on important management issues: delivering products people need, developing processes, working in organizations, providing leadership, and getting people committed.

## Management Theory

The field of management has many interesting facets; and universities, companies, and agencies around the world give much attention to studying and improving it. In his book *Management: Tasks, Responsibilities, Practices*, respected author Peter Drucker showed how the application of management has been the main factor behind the ability to run today's complex society, including both public sector and private organizations (Drucker 1976).

An understanding of management theory involves the simple concept that management tasks (or stages) include planning, organizing, directing, and controlling. These tasks involve anticipating what must be done and making plans, assembling the resources and team needed to complete the work, orchestrating the solution process, and sustaining it after the solution has been developed. For engineers, these tasks often correspond to planning, designing, constructing, and operating a structure, facility, or system. The terms oversimplify the issue, however, because there is much more to civil engineering management: regulatory frameworks, human resources (HR) issues, and legal issues, to name just a few.

The contexts of management situations faced by civil engineers frame the skills needed. Some situations require special knowledge and experience, some require people skills, some require initiative and leadership, and still others require analytic skills. Many engineers eventually face situations that require unusual levels of courage and sacrifice. One of the reasons management is so interesting and challenging is that has so many variations.

A key aspect of management is how people are motivated. Perhaps the most famous theory is that of psychologist Abraham Maslow, who introduced the hierarchy of human needs (Table 6-1; Richman and Farmer 1975).

Another aspect of motivation is the concept of theory X and theory Y. Theory X, an authoritative approach, states that people inherently need to be pushed. Theory Y, which is more participatory in nature, states that people do not need close supervision.

Management is an input–output process. Input includes labor, energy, capital, land, equipment, services, information, and other necessary resources. The organization and everything it does (the production system) is then the transformation process that produces the output of goods and services (Fig. 6-2).

## Management and Leadership

The difference between management and leadership often arises as a discussion topic. When confronted with difficult problems, most people

**Table 6-1**     **Maslow's hierarchy of human needs.**

| Levels Of Human Need | Examples |
|---|---|
| Self-actualization | Achieving life purpose |
| Self-esteem | Being respected by others |
| Belonging and social needs | Love and companionship |
| Safety needs | Physical security and health |
| Physiologic needs | Food, water, etc |

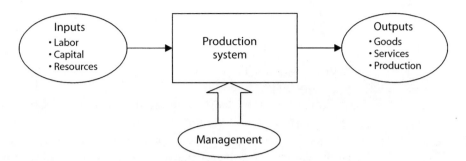

**Figure 6-2**   The production system.

believe that strong leadership is needed. Leadership is an essential quality in all endeavors; without it, not much happens. There are many shades of leadership, however, and different styles are appropriate for different situations. The difference between leadership and management is that leaders inspire followers to achieve objectives, whereas managers allocate resources and carry out other management functions without necessarily getting people to follow (Creech 1994). In other words, leaders get people to follow, and managers run things with authority.

Leadership is not a topic that can be taught easily. People recognize leadership qualities, and studying them can help managers make a leadership self-assessment. For example, leadership qualities taught at the USMA include integrity, knowledge, courage, decisiveness, dependability, initiative, tact, justice, enthusiasm, bearing, endurance, and unselfishness. Other qualities might include fluency of speech, dominance, emotional balance, creativity, self-confidence, achievement drive, drive for responsibility, interest in work, and sociability (USMA 1957).

Leadership is an incredibly important quality in getting things done, and it comes in many forms. Regardless of their positions, civil engineers can exercise leadership by developing a valid vision to accomplish something worthwhile and taking the initiative to get it done.

## Planning as a Management Task

Planning is an important management-related activity for civil engineers. Every phase of management must be planned, including construction and development of facilities, development of management systems, and operation of facilities.

Each management scenario (e.g., projects, businesses, agency programs, management systems, problem-solving processes) faced by engineers requires planning. Plans to be developed include policy plans, capital development plans, project plans, business strategies and plans, financing plans, organizational plans, and plans to improve operations.

These are the front-end development processes that lead to projects, processes, or management systems. Planning is a simple concept aimed at finding ways to get you where you want to go and making decisions about what to do. There are many types of plans, however, and planning jargon can be confusing.

Consulting engineers and public works managers are most concerned with capital development planning for new facilities. Plans might include a master or integrated plan (sometimes called the comprehensive general plan), master plans for each service or type of facility, and needs assessments for the budget process. Plans to improve operations and services will also be needed, as well as plans for the three Rs. Internally, plans to develop the infrastructure management organization are required. This includes facility planning, operations and maintenance planning, and organizational planning. Planning relates to the budget process through the planning–programming–budgeting system. Program evaluation is the final phase of the system. The link between facilities, operations, and financial planning is in the information system.

Organizations such as the American Society of Planning Officials, the American Institute of Planners, and the ICMA publish literature on planning as it relates to community and infrastructure development. The ASCE has an Urban Planning and Development Division.

Businesses and organizations provide another field for planning to occur, as preparing a business plan or a strategic plan is a common requirement to initiate or energize business operations.

## *Planning Process*

The planning process consists of defining goals, developing alternative solutions, selecting the best solution, and putting it into action. This process has many similarities to the problem-solving process and the engineering method. At its root it deals with the desire to accomplish something. Most planning applications for civil engineers fit on a four-dimensional matrix of stages of development, functional divisions of organizations, levels of management, and categories of system. Management stages include planning, organizing, design, and operation. Functional subdivisions of organizations include engineering, finance, operations, and so on. Levels within organizations include the operator, engineer and manager, director, and policy governor. Types of system include transportation, water, energy, and others.

Identifying the problem is the first step in planning and involves filtering information to determine the root issue. This is often the most difficult step because the root problem of a complex issue is often not apparent. There may even be complex, interacting, root causes. The second step and a key part of the planning process is goal setting, which is complex due to the value sets of the many players involved. This can lead to frustration for managers who want quick fixes. Goals cannot be neatly defined by planners in a vacuum; rather they must be determined in a human process. In public venues, it may not be possible to determine goals without an election.

Formulating alternative solutions is the creative step of planning. This step can never be turned over to computers because creativity extends beyond mere enumeration of obvious alternatives. There are technical, financial, organizational, and management alternatives. The evaluation of alternatives is a scientific process involving systems analysis, economics, impact analysis, and political awareness. This is the stage in which computers can be most useful to find cost:benefit ratios, financial payouts, and impacts of alternatives.

The selection of preferred alternative requires decision making. Decision making is a dynamic process because implementation of civil engineering projects in the public arena involves many issues, and decision makers must consider many alternatives and points of view. Studying alternatives involves ranking and presenting them to the decision maker but is not decision making itself. Civil engineering planning can go on for a long time without a decision because it is often unclear, with so many public issues at stake, who has the authority to make a decision, especially when permitting and financing are required and even elections or settlement of court cases is necessary.

As part of the planning process, the effectiveness of a project or management program should be assessed after the fact (sometimes called program evaluation). This rarely occurs in practice, but if the evaluation is valid, improvements can be made the next time. This should be part of any valid quality management program and is part of scientific management.

As complex as all these steps might seem, they only mark the beginning of the planning process. Although the steps outlined show a somewhat linear process, the real world changes quickly and contains a lot of feedback loops. In other words, things do not stand still while you go through the planning process. One demonstrable way that things do not stand still is seen in the people-related dimension of planning and problem solving. Negotiation is an integral part of planning and problem solving. The linear process of problem solving is to some extent sterile in that it lacks a human dimension. People are very much involved in deciding things, and negotiation is a way to transact tradeoffs between them, leading to a positive outcome. It is often best to avoid positional bargaining and move toward a system that focuses on interests rather than positions. Then, by inventing options for mutual gain, objective criteria can be introduced to compare options and devise win–win solutions (Fisher and Ury 1981). The link between the negotiation and problem-solving processes is in dealing with the human factor as the steps of the decision-making process are followed. Each step can involve negotiations, as there may be disagreement even on the statement of the problem.

## Policy Analysis

Policy analysis is concerned with finding the right policies and is the aspect of planning concerned with appropriately steering big decisions. The term *analysis* means to divide something into its component parts; it is the opposite of synthesis, which means to combine components. There are many uses of the word *analysis*, including mathematic analysis, chemical analysis, engineering analysis, and now, policy analysis.

Policy analysis is application of the problem-solving process to finding the best policies to implement. This process requires that problems addressed and possible policies are separated into distinct elements. Policies are courses of action in relation to particular issues. In a legal sense, they have a position in the hierarchy of rules and regulations. A company policy is a rule lacking the force of law but is still an important guideline. The field of social science has a large subdivision called *policy sciences* that covers government and public matters.

In infrastructure, policy analysis refers to the analysis that is done to find lines of action for broad issues. Infrastructure questions that require policy analysis include the following:

- Should the water supply be found independently or in concert with a regional agency?
- Should the solid waste utility be privatized?
- How should the capital program be financed?
- What should be the strategies employed to solve the community's mass transit problem?

The ICMA guidebook on policy analysis in local government presents policy analysis as a systems approach to decision making. There are four essential features: the systems approach, use of the scientific method, use of mixed teams (interdisciplinary approach), and an action orientation (Kraemer 1973). In this sense, policy analysis might be thought of as a variation of systems analysis.

## Risk Management

Risk management is a skill required by civil engineers. Risk management skills are needed in assessing the risk of a particular flood, drought, or wind damage scenario; in assessing the risk of a particular type of failure; or in assigning a probability to any type of outcome, such as being sued for nonperformance or some other scenario. The development of risk analysis as a field was accelerated by disasters such as the Teton Dam failure, which occurred in Idaho and was ushered in with a report by the US National Research Council (US National Research Council 1982). Risks evident today (e.g., terrorism, biological attacks, Internet sabotage) could not have been imagined a few decades ago.

Whatever the risk, it is the job of engineers to temper it by margins of safety, design factors, insurance programs, performance bonds, and other instruments for risk management. These require knowledge about many types of risks and ways to mitigate them. Risk to engineers is often seen as a matter of statistics, and indeed the field requires probability and statistics to provide a quantitative basis for it. Risk analysis involves much more than pure statistics. Civil engineers often find themselves working with insurance companies, bond houses, or other providers of risk management instruments. Engineers working in private practice find they must carry liability insurance to protect themselves against financial loss.

## Organizational Theory

### Establishing Organizations

The right organization is needed to deliver a product or service (Creech 1994); and although the right organization is not a sufficient condition for success, it is necessary for success. It is people—not organizations—who get things done; however, people are not able to work effectively without organization and coordination. Organizational theory is a social science, taught in schools of business and industrial psychology. It considers the interaction of organizational structure, people, and psychology, with communication and coordination being key concepts.

The first task in organizing is to design an organizational structure to carry out the purpose of the business, agency, or unit. The structure should be designed to communicate the purpose to those working inside and outside the organization and should specify leadership, authority, subdivisions of responsibility, and communication patterns. Responsibility and accountability generally flow upward in the organization, but some responsibility flows to workers as well; and although responsibility cannot really be delegated, it can be shared.

### Structure of Organizations

The organization will have titles such as department, division, section, branch, or unit. Each management unit needs a director or manager, and the individual positions within the organization must have job descriptions. Each unit of the organization plans its work with mission statements, goals, objectives, and unit plans and budgets. Accountability is based on those plans, and this is the essence of work planning and management systems.

The center of the organization comprises the units that carry out its direct mission. This part of the organization is normally called line (or operations), and the part that supports it is staff. The line develops the product or delivers the service. In a solid waste collection unit, for example, the line organization would collect the waste and the staff would provide support for the line operation by giving financial management, administrative support, logistics, engineering, legal counsel, and planning. Other functions (such as joint operations or interorganizational coordination) may also be needed.

Organizations have four basic levels: policy, executive, middle management, and worker. These levels recur under different titles depending on the organization. For example, in a water organization, these levels will be referred to as board, director, manager, and operator, respectively.

Variations of organizational types include line, staff, and matrix organizations. Other variations include functional organizations and programmatic organizations. A good organizational structure will release the capabilities of employees and managers for maximum individual performance, minimize the need for realignments and reorganizations, serve as a solid foundation for professional development, and promote effective performance of the organization. Knowledge of the principles for structuring organizations evolved over the past century from such management scientists as Frederick Taylor. Unity of command in organizational design means that each manager or worker should report, as far as possible, to only one supervisor (Sheeran 1976).

A corollary is the span of control principle, which holds that each supervisor should be responsible for no more than a fixed number of persons. The number seven is often used, but opinions vary. Because a main function of supervision is to transmit information, the span of control principle is sensitive to communication technologies and with computers span of control allows more workers to be supervised. This results in a reduction in the need for middle managers.

Another principle is that structure should be goal oriented according to the basic mission. Organizational units should be adapted to the strategic plan of the mission. Units that have no clear identification with the basic mission become candidates for abolishment. Another basic principle is that the organization should be consistently structured and ordered. This means that the geometric layout ought to be balanced, the nomenclature consistent, and the assignment of functions homogeneous. This principle aids in the understanding and clarity of the organization and the avoidance of misunderstandings and confusion.

Finally, the organization should be lean without excess clutter, unnecessary staff, excessive management layers, confusion in reporting lines, duplication of effort or competing units, or artificial barriers to communication. The principles of unity of command and leanness will ensure that organizations have paths of communications, control, command, and intelligence that are straight and unencumbered by many levels of staff, with direct and uncluttered reporting lines. In the real world, of course, some back-channel communication does occur.

The effectiveness of organizations depends on behavioral factors more than on the structure itself. No organization can be more effective than its key employees, and reorganization is common. Industrial psychology explains such factors as motivation, design of work, influence and power, communications, decision processes, performance evaluation, and other concerns that explain these phenomena. The need for coordination and communication within complex organizations has created a need for better communication and has enabled teamwork by breaking down barriers. The matrix organization aims to do this where each employee may report to two or more people. The split reporting illustrates one of the disadvantages of the matrix-organization concept; that is, that it violates the principle of unity of command.

The HR department of the organization includes the kinds and levels of skills necessary to make it work smoothly. Job titles, position descriptions, and work plans for each position are required. The setting of objectives and the design of the work for each position should be set to enable each worker to help meet the organizational purpose and objectives. This system has various names, but it boils down to a general work management system that is a key part of scientific management.

In all organizations workers and managers are engaged in either of two basic activities: doing or deciding what needs to be done. In some ways this distinguishes managers and workers, but it is actually more complex, especially in the Information Age. Organizations will change a great deal in the future, especially because of technologic advances. Today, much attention is being given to making organizations work better. For example, in today's competitive society, the concept of the learning organization has arisen to explain that organizations, like individuals, can learn by experience and become better and more competitive. Peter Senge's *The Fifth Discipline: The Art and Practice of the Learning Organization* explains the basic concept in terms of five disciplines: shared vision, personal mastery, mental models, team learning, and systems thinking (Senge 1990).

## Public Sector Management

Most civil engineers work in one way or another with public organizations. Administering these organizations is the subject of the academic field of public administration. The American Society for Public Administration offers a view of this field, which seeks to advance the art of managing public organizations by focusing on topics such as political science, administrative law, decision making in government, and other issues of public sector concern.

Civil engineers can be involved in public administration in roles such as city engineer, public works director, and city manager. At the scale of large cities and systems, civil engineers have the opportunity to work on such large-scale challenges as regional transportation, airports, megaprojects, and regional environmental problems. Two organizations that can be helpful to civil engineers as they study public sector management are the American Public Works Association and the International City and County Management Association.

## *Business Organization*

Because civil engineers working as consultants are working in private practice, they need the same kinds of management skills that a doctor, lawyer, accountant, or other business owner would need. These include organizing, marketing, staffing, accounting, and other business skills. Civil engineers seeking to start a business need a concept, strategic plan, business plan, incorporation documents, financial plan, staffing plan, and facilities plan. If the firm is successful, it might encounter more advanced issues, such as offers for mergers and acquisitions. Engineers sometimes serve as directors in business organizations, either their own or an unrelated company. The role of director carries with it policy responsibilities that are different from those of management.

# Communication

Communication in organizations is usually more important than how they are structured. Although communication is discussed in more depth in Chapter 8, it is important to mention here how communication impacts organizations. Communication (and other related behavioral issues) is critical to successful organizations. The military's concept for making organizations work—command, control, communications, and intelligence (C3I)—illustrates the importance of communication. The distribution of power in an organization can be traced from communications patterns. Sometimes the organization works differently from the chart or wiring diagram. If communications in an organization are healthy, workers and managers produce better. If not, there is potential for trouble. Good communication is not enough to produce effective organizations; planning, effective work, and executive action are also needed.

Communication within organizations begins between individuals on a distributed basis. These communications can take many forms, including electronic, but they cannot all be subsumed in an electronic office. Written and spoken communications should also be incorporated into organizations. Spoken communications often save time and can be interpreted more clearly with the ability to question, interpret, and clarify. Skill in written communication is especially necessary in complex situations. Managers must master the use of such written communications as memos, letters, reports, analyses, proposals, and press releases. Written communications can be saved for reference, shared with others, and amplified to build successively on the original communication.

One of the fruits of effective communication is coordination of operations. The word *coordination* means to harmonize, and common sense indicates that an organization that is not harmonious will not function effectively. Coordination problems can include poor intraorganizational coordination, lack of contacts between persons working on similar projects, conflict between sections of organizations, and problems with the public. The director of an infrastructure organization should, of course, ensure that all actions are fully coordinated with the higher management and the governing board.

Ways to improve coordination in organizations include use of committees, staff meetings, clarification of coordination responsibilities, appointment of assistants and deputies with responsibility for coordination, use of job descriptions to pinpoint coordination responsibilities, encouragement of cross-communication through a healthy organizational climate, and use of special project-oriented organizational units (matrix approach). With improved decision support systems, including information systems, communication within organizations should improve. Uses for the computer in organizations cluster in three categories: office automation, decision support, and automatic control of operations (Sheeran 1976).

The challenge is to handle unstructured management problems in which flow charts and logic are not applicable. Routine tasks are handled well by computers, so work is effectively pushed up in the organization chart, and the workers who were performing routine work will have to do their bosses' jobs in order to continue their growth. The way to make management more productive is to measure the value added by each function.

# Decision Making

Managers must be able to make effective and timely decisions. When decision making is effective the organization expects quick and effective action. Otherwise, morale suffers and rumors feed on uncertainty. The decision-making process is actually about the same as the problem-solving process. Professional decision making involves both making decisions (cognitive process) and caring about them (affective process). Decision-making skills are vital to critical thinking and communication and show how they involve a sequence of creative and critical thinking processes, which involve analysis, synthesis, and evaluation as modes of thinking. The sequence is applied to five thinking operations (Wales et al 1986):

- Define the situation.
- State the goal.
- Generate ideas.
- Prepare the plan.
- Take action.

The expanded process falls into the categories of analysis, synthesis, and evaluation and consists of a 7-step version of the problem-solving process:

- Identify the situation.
- Create goal options.
- Select the goal.
- Identify goal problems.
- Create idea options.
- Select ideas.
- Identify new situation problems.

Defining the situation involves answering who, what, when, where, why, and how—the elements of case study analysis. Decision making is an art and science, especially unstructured decision making that accompanies such political issues as infrastructure. Science in decision making may use techniques such as decision theory, statistics, and operations research, even when final decisions are based on judgment.

The decision process has five elements: clear realization that the problem is generic and can be solved only through a decision that establishes a rule; definition of the specifications of the solution, or the boundary conditions; derivation of a solution that is right (i.e., one that fully satisfies the specifications before attention is given to the concessions needed to make the decision acceptable); building into the decision of the action to carry it out; and feedback that tests the validity and effectiveness of the decision against the actual course of events. There are a number of techniques to improve the knowledge content of decisions, such as nonmathematic techniques, PERT analysis, force-field analysis, decision trees, utility theory, and probability and statistics (Moody 1983).

Decisions can be classified by level (executive, management, worker), stage of management (planning, organizing, controlling), and function (water supply/distribution, transportation/traffic control, energy/distribution). Decisions are also political, management, or operational in nature. The political level is subsumed in the policy level of decision making, which oversees the executive manager. Managers get into political questions at other levels, although the political content should be minimized.

Decisions are made to solve problems. If the problem is not clear, it should be studied until it becomes clear or more structured. Complex problems should be taken apart to create smaller, less complicated structures that are dealt with more easily. Decisions have different content. The factual content of decisions is relatively nonnegotiable, whereas the value content requires tradeoffs and political analysis. These distinctions are typical of policy versus operating decisions. The operating decision, as opposed to the policy decision, is subject to more definable rules, usually on a shorter time span, more repetitive, simpler, less risky and uncertain, and based more on knowledge of operational data.

Unstructured problems require more experience, judgment, and analysis than do structured problems. The reason computers alone do not result in greater advances is that many of the problems faced in business and government are unstructured. A bias toward action is one of the characteristics of success in decision making.

## *Decision Support Systems*

The use of information to support decisions is a specific task of the organization. The most strategically important decisions take place at upper levels, normally by white-collar workers. One of the challenges is improving productivity in organizations and the use of time by managers. The challenge is summed up by white-collar productivity. Lack of structure in white-collar work impedes measurement of productivity.

To make decisions, managers need information and analysis of the situations they face. Decision processes connect parts of organizations because decisions affect levels and parts of organizations with dissimilar functions but with needs for the same information. The decision support system (DSS) is important to this process. DSS refers to the use of computers to develop and display information to improve decisions. DSS implies more than the processing of data; it includes analysis to add value to data. The main points of the DSS are shown in Figure 6-3, which illustrates the flow of requests for decision support from the decision maker to the staff. There are two main activities in the DSS, data management and studying alternatives, and these are the activities that convert data or information into knowledge that is useful in the decision-making process. DSS is not different from the traditional staff activity of making recommendations to the boss; instead, it is a way to organize the concepts better and arrange to handle information more effectively through the use of computers.

In the future, computers will be used to take over more structured work. One of the vehicles for this is the expert system, which consists of facts and heuristics as well as rules of thumb and know-how. There are three parts to the expert system: a knowledge base of facts and heuristics, an inference procedure to use the knowledge base for problem solving, and a working memory. An evolving research field called "knowledge engineering" is used to prime expert systems. Experts are interviewed in order to glean what they know about a particular problem.

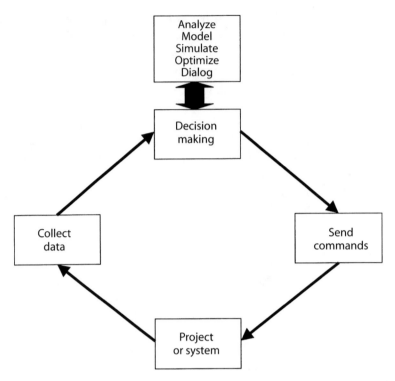

**Figure 6-3**  Decision support system (DSS).

## Strategic Planning

Strategic planning is a powerful and popular method to develop strategy for organizations of all kinds. It involves a structured method and process to identify directions and prepare a plan of action to move toward them. It can be applied in the corporate arena, in military or government organizations, for private volunteer organizations, or for personal planning. A strategic plan will outline goals, objectives, measures of achievement, plans of action, needs and resources, and development plans (Fogg 1994). Strategic planning involves a sequence of decisions: vision, mission statement, and customer identification and understanding. One way to organize a strategic plan is the SWOT analysis (strengths, weaknesses, opportunities, and threats).

## Human Resources Management

Human resources management is key in the management field. Once known as personnel management, the HR field has expanded greatly in recent years to include professional development, employment and workplace law, and employee relations. HR deals with such work force issues as skills, diversity, demand for trained workers, supply issues, work force economics, and improved workplaces.

On the management side, HR deals with work planning by offering systems to break work into tasks. Tasks are organized according to functions within the organization, and this avoids duplication and confusion. If successfully carried out, jobs lead to accomplishment of the overall work and achievement of the organizational mission. The performance and effectiveness of key employees determine success.

The design and evaluation of work in jobs is planned and evaluated according to a fixed cycle (e.g., annually). These approaches are known collectively as management by objectives. Objectives for an individual job or unit should be negotiated and set by discussion and planning. This can be done more often than annually, but the cycle for pay increases and other incentives is normally once a year to connect to the budget. After objectives are set, tasks can be assigned. In organizations, conflict sometimes arises over assignment of tasks. For example, employees in a workplace where they are thought to be lazy are not motivated due to inadequate task assignment.

After plans are made and tasks assigned, coaching and motivation by supervisors are in order. Managers who stay in their offices and do not mingle with workers are neglecting part of their responsibilities. Evaluation is necessary for everyone. Even the chief executive of an organization is evaluated by a level of authority (e.g., board of directors, stockholders). Even if directors are not able to evaluate the executive's work on a day-to-day basis, they can evaluate by results. Stockholders make such an evaluation by how profitable a company is. This same principle of accountability applies to public organizations. The difference is how the profitability is measured and how the oversight group exercises its responsibility.

Evaluation is followed by rewards, increases in pay, promotion, and recognition. Then it is time to set new goals. This cycle of planning, coaching, evaluating, and recognizing is fundamental to HR management.

Organizations involve relationships among people, and the field of HR seeks to make them better. This involves relationships between coworkers as well as worker–supervisor relationships. When conflict arises between workers, supervisors can often detect signs of ill health in the organization. This requires careful and professional attention to employees.

Legal issues in management have increased in number and intensity. Employee and workplace law has become a specialty area among attorneys. They deal with hiring, firing, compensation, harassment, and many other such issues. One of the big areas of HR is equal opportunity. Some categories of people, mainly based on race, gender, age, and disability, have suffered discrimination in the past, and the goal of equal opportunity law is to correct HR practices that need attention.

Intergenerational work that takes place today in organizations requires adjustments between the age groups, including veterans (born before 1943), Baby Boomers (born 1943 to 1960), Generation Xers (born 1960 to 1980), and Nexters (born after 1980). These various generations have differing life experiences that shape their views of values such as authority and loyalty and determine how one works. New methods of communication and a flexible working environment are required to enable these generations to work together effectively (Zemke et al 2000).

## Teamwork

Many professionals, including civil engineers, often find themselves working in teams. Teamwork requires different skills than working alone. For one thing, team members may come from different offices, disciplines, or even generations. They may not like each other or may have other conflicts. Because they are part of a team, however, they share a common goal.

Teamwork requires that people work together toward a shared goal and direct individual efforts toward organizational objectives. The logical analogy is the sports team. On a football team, for example, there is a game, players, a field, coaches, equipment, uniforms, trainers, referees, rules, fans, schedules, practices, team positions, team captains, cheerleaders, leagues, opponents, playoffs, championships, letters, and other sports institutions. These can be translated into the framework of teams of workers attacking a project. Team members must work together rather than as individuals.

It is up to the team leader to clarify to the team what the expectations of each member are. Team members must be held accountable for their actions. One of the problems of teams is that they sometimes have a tendency to conform. To minimize this, a balance must be struck between teamwork and individuality.

## Meetings

One of reasons for business meetings is to work together as a team. Meetings should have a definite purpose and be run well so that people's time is used productively. There are, of course, different types of meetings, such as for coordination, for decision making, and for reporting. A few guidelines on effective meetings are as follows:

- Set a goal and have an agenda.
- Include presentations to summarize known information.
- Start on time and conclude crisply.
- Have a facilitator or chair (not always the boss).

- Have someone record the results.
- End with meeting results and after-action plans.

In addition, there are many other guidelines regarding the dynamics of meetings. Getting participation, keeping a balance among participants, reaching a consensus, and other such goals are important for effective meetings.

## Project Management

One of the first assignments for many civil engineers is to manage a project. Project management involves using methods and tools to work toward a set of objectives uniquely defined in a specific organizational context. A project is an activity with a definite beginning and end and has goals, budgets, and deadlines Projects provide engineers with an opportunity to apply management skills that can produce visible results.

Construction projects are the most common type that civil engineers work on. Other types of projects include organizing an event, preparing a proposal, and undertaking other activities that have beginnings and endings. The discipline of project management is aimed at providing tools to manage these activities. Elements of project management include the following (Lock 1987):

- Participants (consultants, contractors, owners, teams, clients, and regulators)
- Tools (specifications, schedules, plans, task or assignment, displays, inspections, tracking systems, cost estimates and controls, quality assurance, and integrated management systems)
- Management instruments (contracts, accountability arrangements, insurance)
- Legal instruments
- Financing tools (purchasing and credit)
- Integrated management system

Construction is the industry that adds the most to the nation's capital investment. Unless construction projects are completed with quality results and within budget, there can be no solutions to the problems of managing infrastructure. The process for developing construction projects involves planning and designing activities that culminate in construction and completion of a project. Many design–construct activities involve such management tasks as conceptual and preliminary planning, preliminary and final design, preparation of construction documents, contracting, construction, and inspection.

In public works organizations maintaining an engineering department is usually one of the staff functions. This might be called the city engineer's office, or in the military, the post engineer's office. These terms are synonyms for public works directors, and the roles that are reserved for the engineering staff are more distinct. Project development is often handled with a special engineering staff using outside consultants and contractors. After the project is completed it is turned over to the operating staff.

Maintaining the quality of construction is critical to ensuring that investments in infrastructure pay off. There is no more serious breach of the public trust than for a responsible public works official to accept shoddy construction work or to engage in corrupt practices, yet this is one of the problems facing the infrastructure field. Project management is complex and has led many engineer–managers to consider the special requirements needed to succeed in managing projects.

Construction is sometimes regarded as an engineering function, but many skilled construction managers are not engineers; in fact, there are some very effective academic programs in place to train such managers. The American Council for Construction Education lists the academic programs at which construction management can be studied. These include engineering colleges, technology colleges, schools of architecture, and such other units as the Colorado State University Department of Manufacturing Technology and Construction Management. Obviously, management is a function that can be approached by different disciplines.

## Operations Management

After a project is built, it must be operated and maintained. This requires operations management, which means to focus on the operational strategy of any business or organization. Operations management includes production management, facilities management, maintenance management, and information management. When the organization is operating, it is fulfilling its basic mission. In public service organizations this guards against straying from fundamental purpose and becoming ossified with bureaucracy; in private sector organizations, it keeps the focus on the bottom line—the essential elements of survival and growth. The following are important operational areas for a well-functioning organization:

- A clearly stated organizational purpose
- A valid and effective organizational structure, with good communication and information flow
- Operational missions and objectives for the subunits of the organization
- Job descriptions throughout the hierarchy of the organization
- Plans and production targets for the subunits of the organization
- A work management system
- A method for checking performance
- Procedures for making changes

Operations improvement has led to the field of scientific management, which was brought to the industrial world in the late nineteenth and early twentieth centuries by persons such as Frederick Taylor. Taylor was a famous engineer who became an efficiency expert and was a pioneer of industrial engineering, the branch of engineering concerned with measuring work and making it more efficient. In the early days of scientific management there was contention between owners and labor; this contention led to some of the labor movements that sought to provide for greater quality of work for laborers and less exploitation by owners. Vladimir Lenin, a father of Soviet communism, referred to the measurement of work in the scientific sense as inhuman, treating workers as animals (Grigg 1988). These facts come to bear on the reasons there is still resistance in implementing such programs as effectiveness measurement, especially in government.

In operations management, there is a system to be operated with certain objectives. The system can be an infrastructure service, such as transportation, water, buildings, or solid waste, or it could be other types of services such as a bank or restaurant. The system is managed by the control system, monitored by the data collection system, and decisions are made in the headquarters with the aid of the decision support system. The essence of the items is captured by the military term *C3I*, which provides for decision making and the issuance of control orders. Control means the capability to actually manipulate the system, such as through a functioning organization equipped with the right control devices. Communications includes all data flow, telecommunications, written orders, and other forms of communication necessary to operate the system. Intelligence is the collection of data necessary for management and decision making.

Productivity is a tool for evaluating how well an organization is doing. It measures the output resulting from such input factors as work units, materials, energy, and cost. An example can be seen in public transit. The overall objective of public transit is to maintain a system that provides access to places where citizens want to go in a safe, quick, comfortable, pleasant, convenient, and reliable manner and that helps minimize pollution, congestion, and energy consumption in the community. Each of the emphasized characteristics becomes an objective (e.g., quickness is measured by rapid movement of an objective). The main quality characteristic of this is travel times, measured by actual travel times from the schedules. To keep track of this requires data of actual schedule compliance (Grigg 1988).

The pertinent questions in production are these: What is produced? What are the quantity and quality of the output? What is the cost, and what are the external impacts? These questions are easier to identify when the operation is a private business and the products are sold in the marketplace, but when the product is a public service it must be analyzed using different approaches that center around defining the service levels needed and supplied.

Productivity measurement for infrastructure categories would generally use such parameters as cost per 1000 gallons of water delivered, tons of solid waste collected per work-hour, cost per mile of street cleaned, cost to maintain a certain amount of the street network, and related parameters.

Effectiveness means doing the right thing efficiently. Efficiency alone is not enough to evaluate an enterprise because it does not test whether the right goals are being pursued. The bottom line of productivity and program evaluation is to ensure that the operation is performing effectively. Effectiveness measures for infrastructure services allow analysts to consider their use in program evaluation and related planning. An Urban Institute report lists some 11 types of performance measures (Hatry and Peterson 1984):

- Cost
- Workload accomplished
- Effectiveness/quality
- Efficiency/productivity
- Cost:workload ratios
- Efficiency/effectiveness
- Resource utilization
- Productivity indexes
- Pseudomeasures
- Cost:benefit ratios
- Comprehensive performance evaluation

Management audits for operations ask questions such as what are the goals of the operation, how can the success of the enterprise be measured, and what are the numeric parameters. Ideally, performance of the infrastructure systems should be audited regularly.

Maintenance is a key support activity for operations and requires an MMS to determine whether the system is working as it should. As discussed in Chapter 5, maintenance involves four separate functions: condition assessment, inventory, preventive maintenance, and corrective maintenance. The condition assessment activity is a link between the operation and the maintenance function and illustrates why the two functions must be unified. Preventive maintenance is intended to head off problems. Corrective maintenance involves repair, replacement, and rehabilitation of facilities.

Operations and maintenance have different but related information needs. Information related to operations subdivides into the part that deals with facilities (hardware) and the part that deals with workers (software). Operations managers need information and reports on how well the systems are doing. The logical extension of operational information systems is movement toward automatic control. Maintenance information should support MMSs and provide information for budgeting as well as management. There are a number of commercial software packages for maintenance management.

In public works organizations the engineering function is responsible for the studies and designs necessary to construct and reconstruct facilities. The engineering department would normally be responsible for numerous records, especially as-built drawings of newly constructed or reconstructed facilities. Also, any system maps would be the responsibility of the engineering department. There is an obvious link with the information needs of operations and maintenance, and a common database should be available.

# Quality Management

Quality management and control are concepts that have become important to industry, which has learned that high product quality is essential and that the best way to have high quality is to do things right the first time. The concept of QC has expanded into a wider view of quality in all aspects of management (that is, having it improve continually). QC means to ensure that quality of product, whatever it is, is within acceptable limits, as defined for that particular product. Sometimes QA is used in place of QC.

Quality control functions are easy to envision in a water treatment plant that must meet regulatory requirements. If the goal of the utility is to deliver service that involves water of certain quality, with certain pressures, and during certain time periods, it would need an inspection program, a quality control sampling effort, laboratory work, and record keeping. Another example might deal with storm water. If the goal is to prevent a certain level of flooding, the indicator might be the absence of complaints but the organization could have an assessment program.

There is much to learn about management from quality improvement. One version is called total quality management. Bill Creech, authority on management, wrote this about total quality management: "Product is the focal point for organizational purpose and achievement. Quality in the product is impossible without quality in the process. Quality in the process is impossible without the right organization. The right organization is meaningless without the proper leadership. Strong, bottom-up commitment is the support pillar for all the rest. Each pillar depends on the other four, and if one is weak all are" (Creech 1994).

W. Edwards Deming became famous for his suggestions about improving quality of businesses through the Deming method, which has been reported in many places. The Deming principles include the following: (1) create consistency of purpose with product and service; and (2) instead of focusing solely on making money, focus on staying in business and providing more jobs through innovation, research, constant improvement, and maintenance. The principles Deming presents apply to all organizations (Grigg 1988):

- Adopt the new philosophy.
- Cease dependence on mass inspection.
- End the practice of awarding business on price tag alone.
- Improve constantly and forever the system of production and service.
- Institute training.
- Institute leadership.
- Drive out fear.
- Break down barriers between staff areas.
- Eliminate slogans, exhortations, and targets for the work force.
- Eliminate numeric quotas.
- Remove barriers to pride of workmanship.
- Institute a vigorous program of education and retraining.
- Take actions to accomplish the transformation.

## Program Assessment and Management Audits

Just as you can have a financial audit, a management or performance audit can identify ways in which management effectiveness can be improved. One of the functions of a board of directors is actually to bring an outside view to an organization. Because they will be evaluated, management and staff should stay on their toes. Management consultants can be used for this task, and a team of them can visit an organization for a short period to assess the organization's purpose, customers, goals, procedures, and results. It is amazing how much insight outsiders can bring to the operation of a program that the insiders understand but have come to take for granted. A management audit is like an annual review of an employee but is applied to an organization. Methods can include visits, interviews, surveys, inspections, document review, financial studies, and other tools available to management to assess performance.

## Facilities Management

Facilities management is an operational phrase that means operation and management of capital facilities, which depend on planning and maintenance. Due to financial constraints, there is interest in making existing facilities last longer and operate better. The need to make infrastructure facilities work better is illustrated by the three Rs programs used by public works agencies around the nation, which focus on repair, rehabilitation, and replacement of facilities. There is also a four-R highway program (resurfacing, rehabilitation, repair, and replacement). These principles may seem elementary, but it is surprising how many engineers involved in public works have not dealt with the important questions of operations and maintenance tools of management.

## Marketing and Public Relations

Marketing and public relations are two key areas in business and management, and while they are primarily used in the for-profit world, they also apply to nonprofit organizations. Even churches use some aspects of marketing and public relations. Marketing includes a spectrum of activities and is not limited to sales alone. It begins with identification of the market and the business's customers and the products or services they would like to have, then proceeds through the sequence of product and service development, pricing, promotion, advertising, selling, and distribution.

Public relations is a closely allied field that includes identification of target audiences, use of media to achieve objectives with them, and work with organizations to achieve objectives. A few of the techniques that are used in public relations are news releases and media relations, feature articles, newsletters, annual reports, special events coordination, celebrity speakers, fliers, invitations to events, and other ad material such as posters and signs.

Until recently, it was considered unethical for professionals (such as lawyers and engineers) to advertise other than a notice in the phone book. Today, some professionals advertise freely, especially some categories of attorneys. Engineers still do not, for the most part, "advertise," but they certainly engage in public relations. For example, a new engineering firm will have brochures, business cards, a web site, newsletters, receptions, and other publications, events, and notices designed to attract the attention of customers.

## Personal Management Skills

Young civil engineers have a lifetime of professional development ahead, with education just the start of personal development. In addition to an education, civil engineers should recognize the importance of developing personal management skills as well as skills that can be used to manage an organization. A number of recent books aim to guide professional people in becoming more effective. These books cover such topics as goal setting, time management, getting a personal vision, negotiation, and interpersonal relations.

Maslow's hierarchy of human needs (Table 6-1) demonstrates how to assess personal management skills. This assessment begins with basic survival needs and ends with self-actualization and personal fulfillment needs. People continue to seek something (e.g., achieving a goal) even after satisfying their basic needs. When people feel fulfilled, they believe they are successfully following the right path for their lives. They also believe they are effective in achieving goals. The key is to have a plan in place and to identify the goals.

Jimmy Carter, in *Why Not the Best?: The First Fifty Years*, discussed his interview with Admiral Rickover, legendary developer of America's nuclear Navy. Rickover asked Carter whether he had done his best at Annapolis. After thinking for a minute, Carter said "No, sir." After the interview, Rickover left Carter with the question, "Why not?" This experience gave Carter something to think about and affected his whole life. What Rickover was really asking was, "Do you know what you are supposed to do, and have you prepared for and determined to do your best to achieve that?" (Carter 1996).

Improving personal management skills involves a whole system of activities: knowing your purpose, clarifying your values, getting your vision right, assessing your strengths, setting goals, thinking and strategizing, getting motivated and committed, taking action, and assessing results. In many ways, personal management is the individual version of strategic planning, and some management gurus have compared the two by suggesting that every individual should have a personal mission statement.

Like other aspects of management, the personal management system has several components, including planning, organizing, and controlling. Planning is the process of setting goals and developing strategies. Every person must apply planning in a unique way. An artist will have less use for personal planning than will a businesswoman who supervises many projects. Even an artist must use time well, however. Developing a plan and having the right goals require a recognition of one's purpose and a personal vision of life. Every person needs a compass to direct his or her life. Understanding one's full potential requires an understanding of individual purpose and skills. Having this compass right is a key point of author Steven Covey in his best-selling *Principle-Centered Leadership* (Covey 1991).

Knowing which direction on the compass to follow requires a personal vision. Vision is the way one sees or conceives how life will unfold, the foundation or root of personal motivation—in effect, a life plan. Unfortunately, many people lack a personal vision, and finding one can be a complex task—especially later in life after a person has missed opportunities to be mentored, advised, and encouraged. There are many sources of confusion and discouragement. It is important not to become confused or discouraged. It is also important to know how to develop personal values and purpose. Techniques that can be used to assess vision and effectiveness include writing an autobiography, writing an epitaph, keeping a running journal, and taking personality tests.

Goals flow from vision. A person must be an organized thinker and planner in order to set goals. Goals must be set for both the long and short term. Obtaining a college degree is a long-term goal; a daily "to-do" list contains short-term goals. Short-term goals should be those that will lead to achievement of long-term goals. Today, personal calendars and software are loaded with tools to help people set goals. Setting goals is not enough, however, without a strategy to achieve them. Developing a strategy involves specific goals and actions. Imagine that part of your vision is to help people achieve their purposes through improved education and your goal is to become a teacher. Depending on your age and education, you will need a college degree in education and a plan to get a job as a teacher. Each of these goals and actions will require a series of steps, which, if followed, will enable you to achieve your goal and fulfill your vision.

Personal management skills can help but not ensure success in goal achievement. Personal skills in planning, organization, time management, and finance are described in numerous books, magazines, audiotapes, and seminars. Sometimes these skills are presented as panaceas in solving problems, but each is simply a tool. Personal management tools are not enough to achieve success because a person also needs a good attitude, commitment, principles, and the will to follow through. These are, to some extent, character traits, which can be learned as well. A positive attitude and good motivation are critical ingredients to success. Getting encouragement and a mentor can help.

## Character Building

Good character traits are essential to success, and this is being recognized in the United States today with renewed emphasis on character education. In 1999, for example, Colorado State University was recognized as one of the country's top universities that encourage character development. Some of the university's character development programs and actions include a commitment to helping students lead ethical and civic-minded lives; a volunteer service program; help for students to strengthen their values, moral reasoning skills, and ethical decision-making skills; and guidance for moving across cultural boundaries and acting ethically in an increasingly complex global society.

## Individual Professional Development Plan

Just as there is a curriculum to follow in college, it is important to have a personal professional development plan. This plan can include goals, vision, and a personal mission statement; associations to join and mentors to work with; a personal assessment plan; learning goals and plans; a personal journal; identification of role models; a reading list; a pocket calendar or personal organizer; a time management system; learning a language or new skill; a deliberate plan to get new experiences; skill-building courses, such as public speaking; a professional portfolio and resume; and exercise, conditioning, and wellness programs.

# Entrepreneurship

At some point in their careers, engineers might begin to yearn for something different (e.g., to start a business, implement some kind of innovation). Some engineers are entrepreneurs. An entrepreneur is someone who is willing to take risks to innovate or get something accomplished. An entrepreneur in business will take risks to start an enterprise. Perhaps the best known and certainly the wealthiest entrepreneur today is Bill Gates, founder of Microsoft. Entrepreneurs can also work in the nonprofit world. For example, the entrepreneur's vision might be to start a new outreach organization to meet a pressing need, such as world hunger.

An innovation is performing some action with a new or better method. Innovators are highly valued in organizations. Suggestion boxes were meant to glean valuable suggestions from people for improvements. Recognizing that technology, computers, and systems analysis have much to offer, there have been many attempts to innovate in public works management in recent years. Sometimes innovation works; sometimes it does not. For example, systems analysis helped us get to the moon, but when introduced into the public sector it elicited negative comments. "If we can land a man on the moon, we can certainly pick up the garbage in New York," meant that it is harder to pick up the garbage in New York than to land on the moon. Although made tongue in cheek, this statement underlines the difficulty of public works in some settings.

During the 1970s, there were public technology programs and cities got public funding for technology agents. Public Technology Inc. was formed to apply technology to cities. Innovation did result, but much of it was private initiative rather than government funding. Government funding, however, can provide incentives for private entrepreneurs.

# Volunteer Organizations, Community Service, and Networking

One of the helpful aspects of management is the opportunity to be involved with leadership figures from the community, professional societies, other businesses and organizations, and nonprofit groups. To become involved with these organizations, it is good to become involved in civic and business activities such as the chamber of commerce, service clubs (e.g., Rotary, Lions, Kiwanis), and professional societies. To the extent that engineers can find the time, they can gain from seeking leadership positions (e.g., committee chair) in organizations such as ASCE, the chamber of commerce, business organizations, and civic groups.

A main advantage of such involvement is the fact that it expands the engineer's circle of contacts. To manage this circle, it is well to have a database of contacts for networking. Periodically contacting the most important people in the database (regardless of whether there is an immediate need) helps to maintain this network.

Beyond the immediate benefits of volunteer work, there are many societal needs that must be met. Philanthropy exists to help to meet these needs, which go far beyond what government can handle. Because of their service orientation, engineers are often drawn to the nonprofit world. Examples of organizations they might support include technical societies, emergency response charities, international development agencies, and environmental organizations. For example, the American Water Works Association helped to spin off the humanitarian nonprofit organization Water For People. This organization is "a nonprofit, charitable organization in the United States and Canada that helps people in developing countries obtain safe drinking water."

# Frontiers of Management: Peter Drucker, Management Guru

Management guru Peter Drucker turned 90 in 1999. Drucker crystallized the discipline of management with his 1954 seminal text *The Practice of Management* (Drucker 1954). Drucker displayed remarkable insight about management and called it the "least understood of social institutions." He understood that if people could learn management, ordinary people could achieve better results and organizations would become part of the fabric of society. To get a perspective on this, consider that before 1900, organizations were not at all prevalent. At that time, 80% of people worked on farms or with their hands, usually working alone and not in teams. Today, knowledge workers, engaged in technical and professional specialties, are the largest group of workers. Drucker showed how management articulates an organization's purpose and translates it into performance. He is famous for asking such questions as "What is our business?," Who is the customer?," and What does the customer value?" This became the theory of the business. Profit indicates whether the theory works. It is the equivalent of the scientific method in business, where hypotheses are tested in action, then revised. Drucker defined the challenge of the twenty-first century as raising the productivity of knowledge workers, including engineers.

Drucker also focused on the nonprofit sector, where he thought that the application of management principles could yield results. While others focused on e-business, Drucker recognized that knowledge workers do not like to be managed; instead, they must be supervised with a focus on performance and results. Knowledge workers need to think about what they are good at, how they learn, what they value, and to obtain the self-knowledge that will lead to effective performance. Drucker helped to identify how the quality of our lives and our society depends on the quality of the organizations that are built and on the quality of the management that runs them (Magretta and Stone 1999).

# 7 Critical Thinking

## Introduction

Critical thinking skills are important to civil engineers and, indeed, to anyone facing significant challenges in the workplace. This chapter explains more about critical thinking skills as well as such associated concepts as creativity, problem solving, and systems thinking. In order to explain the importance of critical thinking, imagine this scenario:

> You have worked for a small city engineering department since receiving your civil engineering degree several years ago. During this time you designed several roads and intersections drawing on the knowledge you gained in university courses (i.e., how to size intersections based on traffic flow and lay out curves based on the speed of traffic). Based on this experience, you believe you have grown a great deal professionally.
>
> One day, your boss says that because of your outstanding work to date you are going to manage the new large project being planned by the city. You are given an overview of the problem the city is facing in providing a truck bypass. The need exists to funnel commercial truck traffic around your city to minimize the negative impact the current traffic is having on your city streets. This is a new type of problem for you in that it is no longer a matter of developing proper curves for roadways or the appropriate subgrade thickness for a road surface. Now you must make decisions about questions that do not have right or wrong answers

but rather have many possible answers. These are typical of the planning and policy questions civil engineers face.

You start by drawing on your education and examining maps for possible route locations based on the terrain, existing roads, and other such factors. You follow this with some preliminary designs to get early cost estimates that can be used to develop a list of potential solutions for further development. You then present your conceptual design to your boss. Once your boss is satisfied with the quality of your work, you must make a presentation to the city. Armed with calculations neatly prepared in a series of computer-generated slides, you go before the city council to present your ideas. Although you are confident, you feel nervous about making your first presentation to this type of audience. You start the presentation and gain confidence as you present your ideas and the numbers with a quick overview of the supporting calculations. After responding to several technical questions, the next question asked is, "Why do all of your routes go through this one sparsely developed section of town?" You give technical advantages that point to this area as being a great location for a highway bypass (e.g., little cutting of rock, easy access to existing roads).

Although these are great technical reasons, the questioner is unsatisfied because his voting constituency lives near this area and will perceive that they will pay a heavy price in terms of noise and air pollution for the benefit of the city's reduced traffic. The questioner's constituents will not like this. The technical background you gained in college and over the past few years can no longer provide the answers you need in this meeting. The question requires you to look much deeper into the choice of road location than technical issues of how best to build a highway. It is no longer a question of "Can it be built?" but instead "Should it be built and where?"

This scenario is being played out in many cities. The world of civil engineering is no longer isolated from today's political, social, and economic problems. Civil engineers are increasingly called on to provide more than technical solutions. Civil engineering solutions must protect the environment and be socially and politically acceptable as well as economically and technically sound. Your career is going to require that you develop the critical thinking skills necessary to deal with these types of issues on a regular basis.

To put critical thinking into context, consider that there are a number of competing alternative solutions to the scenario presented. Each alternative has advantages and disadvantages that look different from one criterion to another and from one affected group to another. The problem is not linear and one-dimensional at all but has at least three dimensions—

alternative solutions, contributions to achievement of goals, and impacts on different groups.

Critical thinking skills can be applied to find workable solutions (i.e., solutions that work best when all things are considered). Critical thinking skills are important for civil engineers. For example, Norman R. Augustine, president of the NAE, wrote that in recent years the subject matter of engineering has changed, but the more profound change is that "political and economic limitations, rather than technical ones, increasingly are decisive in determining what engineers can accomplish" (Augustine 1996).

## Critical Thinking Defined

The roots of critical thinking and problem solving are in the realms of philosophy and rational thinking. Some contemporary thoughts on critical thinking are attributed to John Dewey. Although Dewey did not actually define critical thinking, he put a framework on it by proposing that critical thinking involves bringing closure to uncertain or problematic situations in which there is no way to apply a formula to derive a correct solution and no way to prove definitively that a proposed solution is correct (Dewey 1933).

Educators seek to impart critical thinking skills to students, and according to L. Lipman, a researcher in the field of education, "critical thinking is skillful, responsible thinking that facilitates good judgment because it relies upon criteria, is self-correcting, and is sensitive to context" (Lipman 1988). J. Chaffee, an author who has studied the process of critical thinking, refers to people as critical thinkers if they "have developed thoughtful and well-founded beliefs that guide their choices in every area of their lives" (Chaffee 1998). Everyone should be aware of their biases and be willing to explore situations from perspectives other than their own. In addition, everyone should develop sound reasons to support their points of view.

Two elements common to definitions about critical thinking are considering issues from many perspectives (context) and considering what can be known and how something can be known.

## Perspectives

To examine perspectives, consider again the scenario of the truck bypass and notice especially how viewing the problem from different perspectives might result in a better solution. The first step in critical thinking is to identify our personal perspectives (and possible biases) and then consider other possible perspectives.

Engineers viewed the new design problem of a bypass as a larger version of existing problems (i.e., road widths, curve layouts). This is natural because it is natural to try to understand new problems as extensions of (or in relationship to) problems encountered in the past. This is a good first step in approaching a new situation because it builds on the existing knowledge base.

The logical next step is to identify issues that are different from past experiences. Here it is important to move beyond personal perspectives and to understand, incorporate, and acknowledge other perspectives that may lead to a better understanding of the real problem at hand. During the problem-formulation stage, the opportunity to view the project from the perspectives of all interested parties should not be missed.

Being able to anticipate and understand perspectives other than our own is a difficult but useful skill that requires practice. It is important to be critical of our own work. It is very easy to become satisfied with personal successes. Engineers cannot allow previous successes to become the only reason for future designs. It is also important to consider other perspectives. Civil engineering is a profession that serves society. As such, civil engineers must be aware of the needs, desires, and concerns of society in general. A good starting place is to follow public discussions in the news. If the engineers working on the bypass project had been following public discourse on the truck issue, they would have been aware of neighborhood and political concerns and how interest groups were staked out on the issues.

A second step in understanding different perspectives is to seek out opposing opinions; however, critics also bring to bear their own biases. Much can be learned by understanding what critics say. This can mean both personal critics and more general critics, whose criticisms are often published. For example, the book *Divided Highways: Building the Interstate Highways, Transforming American Life* (Lewis 1997) provides many useful lessons for engineering students involved with future transportation projects. The book presents many perspectives on the development of the interstate highway system and discusses political maneuvers that affected the system both positively and negatively. There is also discussion about the concerns of the people whose homes, neighborhoods, and lives were changed by the highway system. Engineers are well served to be familiar with these historical perspectives on one of the great civil engineering achievements of the twentieth century.

Ultimately, viewing a problem from multiple perspectives helps in the development of a more complete picture of the context of the problem. Critical thinking is sensitive to context. Chaffee provides guidance on how to develop our awareness of context. He suggests that the following questions be addressed: Who? What? When? How much? How urgent? Answering these questions in a critical manner requires that engineers "become aware of your own biases, explore situations from many different perspectives, and develop sound reasons to support your view" (Chaffee 1998).

## *The Question of "Who?"*

Consider the question of "Who?" in relationship to the bypass scenario. To determine who is impacted by the bypass requires looking beyond the immediate physical setting of affected populations. The first answer to the question "Who?" must include people living in the areas where the current traffic is routed and the people in the areas of the proposed routes. However, these are not the only people affected. For example, the truckers who drive the route should also be considered. Broadening engineers' concept of the affected population will result in a solution that is acceptable to more people. This is known as *stakeholder analysis*.

## *The Question of "What?"*

The question of "What?" can also be framed from several different perspectives in the bypass scenario. First, it can be viewed as mainly an issue of removing trucks from the city streets. This point of view can quickly lead to solutions that focus solely on truck traffic. A broader perspective is the need to reduce traffic congestion along the city corridor. If this is the problem statement, then the solution will need to address more constituents than just truck drivers (e.g., automobiles, busses). Another motivation for the project may be increased pollution in the city. Each of these perspectives taken alone can constitute a problem statement. More likely, however, a combination of these perspectives is the real problem to be solved.

One of the easiest mistakes to make is limiting our perspective when developing a problem statement to what appears to be the most obvious problem. If this shortcoming is not detected until a solution is being proposed, the constituent group affected will quickly bring its perspective to the forefront. Although it can be difficult to determine all the different perspectives to consider, engineers can make progress by developing the skill of listening very carefully to the requests of clients and probing deeper into the situation. These skills are part and parcel of problem definition in the planning process (see Chapter 6).

## Knowns and Unknowns

One of the differences between situations faced by both engineering students and practicing engineers is the type of problem being solved. During the early portions of most engineers' careers, problems tend to be structured and to have correct answers. The solution process required to develop the answer may be complex, but answers derived from the process do solve the problems. Engineers are well prepared to solve these structured problems, and much of the engineering curriculum involves them. The more difficult problems deal with making choices in unstructured situations when the circumstances involve uncertainty and there is no single correct solution. More often, many solutions exist, all of which have both positive and negative aspects.

Problems that do not have a clearly defined and unique solution are often referred to as open-ended. Most engineering design problems fit into this category. The design process (see Chapter 5) also deals with problems. Design alternatives must be critically reviewed early in the planning process. Designing is an iterative process that requires feedback between the steps (Dym and Little 2000). Such feedback information comes from the critical analysis of each step. Engineers must apply critical thinking skills continually throughout the design process.

To be effective in providing feedback in the design process, it is important to use evaluation criteria when making the critical analysis of the output from each design stage. Often this criteria must be developed by the design team to test designs as they proceed during intermediate stages. This is in accordance with the definition of critical thinking (Lipman 1988). Most engineering design books provide guidance on criteria for evaluation (Koen 1985, Love 1986, Cross 1994, Dym 1994, Voland 1999, Dym and Little 2000).

Most engineering problems are filled with uncertainties. No matter how precisely data are collected and summarized, uncertainties will exist and call for critical thinking skills. It is not the uncertainty that causes the problem, it is the lack of acknowledging the uncertainty that is dangerous. There are many techniques for handling uncertainty, many of which are included in the undergraduate curriculum. Critically thinking engineers use these tools and base judgment on acknowledgment of uncertainties.

The bypass scenario seems to contain more uncertainties than known data. Data collection will be an ongoing task, starting with the initial stages of problem formulation through the latter stages of final design. There will be many sources of data, some under your control and others provided externally. It is important that both the source and methods of data collection be critically reviewed. In the bypass scenario, for example, data are needed to determine the number, size, and temporal distribution of truck traffic on existing routes. This will require measuring these quantities in the field. If the data collected are to be valid, the measurement period must cover all typical circumstances of traffic. It is easy to make isolated measurements and then substantiate conflicting opinions about the actual traffic problem. Critical thinking requires an understanding of what can and cannot be known and the biases attached to what is known.

Traffic data are not difficult to measure, but trying to predict or measure the environmental impacts of a project might seem impossible. This is often a source of conflicting data. Each source will bring its own bias to the methods of collection and presentation. The job of civil engineers is to recognize situations and evaluate the value of data by not taking it at face value but rather applying critical thinking skills to identify biases and quality of methods.

# Systems Thinking

Critical thinking skills are closely related to those needed in any problem-solving process and are found in techniques such as systems thinking. Problems can be classified in many ways (Table 7-1). The problem-solving process involves a series of steps:

- Problem identification
- Establishing the decision-making process
- Goal setting
- Establishing decision variables and determining how success is measured
- Involving stakeholders and decision makers in the process
- Studying the variables, problem environment, and constraints
- Finding alternative solutions
- Evaluation
- Studying impacts and stakeholders who are impacted

**Table 7-1**     **Classification systems.**

| Classifiers | Examples |
|---|---|
| Issue | Forensics |
| Stage of project | Planning, design, construction, operations |
| Complexity and difficulty | Complex, simple, difficult |
| Arena | Science, politics, business |
| Level | High level, low level |
| Scale | Large scale, small scale |
| Structure | Structured, unstructured, parallel, sequential, iterative |
| Consequences | Inconsequential, risky |
| Knowledge bases | Single discipline, multidisciplinary, interdisciplinary |
| Area of knowledge | Political, social, economic, engineering, biological, mathematic, etc. |
| Persons involved and responsibility | Military, civilian, group, individual, etc. |

- Decision
- Implementation

Bernard H. Rudwick, a systems analyst in the aerospace industry, explained the differences in problem structure: "At one end of the spectrum are the mathematically oriented analysts who wish to apply a set of optimization techniques to highly structured problems. ... On the other end ... are those analysts whose starting point is the unstructured problem of the decision-maker. Their major objective is to build a proper structure to the problem, including uncovering the true goals of the decision-maker" (Rudwick 1973). Unstructured problems tend to migrate upward to the managerial and policy levels. Hard technologies tend to be used by analysts rather than by decision makers.

Civil engineers must function as analysts, who explain the complexity of issues, and as synthesizers, who seek sociotechnical solutions for complex problems of multiple stakeholders. They must use synthesis and analysis to find solutions, and critical thinking is a tool for both. Systems thinking and taking the systems view mean to take different perspectives and apply knowledge correctly through mental models and process modeling. This requires us to measure systemic links and view problems holistically (Senge 1990, Grigg 1996).

Conceptually, systems thinking is a way to apply systematic methodologies and philosophies to the analysis and synthesis of complex problems and organizational issues. Peter Senge, a Professor at MIT's Sloan School of Management, identified five component technologies of the learning organization (Senge 1990): systems thinking (the "fifth discipline"), personal mastery (competence), mental models, building shared values, and team learning. The focus on the organization enables a blending of analysis (systems thinking and mental models) and decision making (building shared values and team learning).

Engineers often use models, which are part of the systems thinking toolbox. For example, systems dynamics models, developed by Jay Forrester of the MIT Sloan School of Management, are sometimes used for simulation of complex systems (Forrester 1961, 1969). It is interesting to note that Senge was a student of Forrester.

For many projects (such as the bypass scenario), the relevance of models depends on how well they can simulate comprehensive issues. Unfortunately, models are more effective for simulating subsystems and mechanistic processes than they are for such complex issues as the bypass. The quantitative aspects of systems thinking mostly involve models of subsystems, and systems thinkers must conceptualize systemic effects.

Another group of critical thinking techniques uses public involvement and integrated resources planning (Priscoli 1989). Public involvement techniques help engineers gain agreement on facts, alternatives, and solutions. They include such techniques as public information, task forces and advisory groups, public meetings, workshops and problem-solving meetings, conferences, mediation sessions, collaborative problem solving, negotiation, and arbitration (see Chapter 8).

Integrated resource planning (IRP) is an important critical thinking technique (Grigg, 1996). IRP is a vehicle designed to assist with the following:

- Defining the overall goals and objectives and establishing milestones
- Identifying all the stakeholders and their concerns and involving them throughout the process
- Determining the problems, critical planning issues, and potential conflicts to be addressed during the process
- Identifying and managing risks and uncertainties
- Implementing the IRP
- Evaluating the effectiveness of the process and making appropriate adjustments

## Lifelong Learning

Civil engineers must continue to grow in their professional knowledge and anticipate how critical thinking skills are needed. Estimates for the useful half-life of an engineer's knowledge are sometimes given to be as short as 4 to 7 years. The implication is that engineers must continually update their level of knowledge and skills. Although no one knows for sure how fast technologic information will continue to grow, there is no dispute about the need for engineers to pursue more and better knowledge throughout their careers.

As discussed in Chapter 4, civil engineers are required to develop their initial skills and knowledge base through a formal education process in an undergraduate degree program. Engineers must also use other avenues to continue their education, such as short courses, conferences, graduate courses and degrees, journals and other professional publications, and on-the-job training. In addition to knowledge per se, continuing education should promote improvements in critical thinking.

# 8 Communication for Civil Engineers

## Introduction

If there is one topic that college graduates agree is important, it is learning to communicate better in all areas of life. This chapter explores communication skills for civil engineers; however, it does not attempt to cover all aspects of communication and is not a how-to chapter on writing or oral presentations. Many excellent books are available on these topics (Beer and McMurray 1999, D'Arcy 1998, Davies 1996, Eisenberg 1992, Michaelson 1990). Communication skills are important to all professionals, not just civil engineers. The goal of this chapter is to provide civil engineers with the motivation to seek additional knowledge that will assist them in becoming better communicators.

As a group, engineers are not known for their communication skills, although many defy this stereotype. Regardless of the setting, the ability to communicate can often mean the difference between success and failure in a project and ultimately in a career. The best-kept secret may be that civil engineers are very effective in public forums, where the dialogue driving the democratic process takes place. As we have learned when meeting with employers, when asked about what they need in engineers, they often list communication skills higher than technical skills. Having good ideas is one part of the equation; being able to express them is the other.

This chapter examines the purpose (why) and the intended audiences (who) of communication but does not cover the methods (how). This chapter reviews the types of communication for civil engineers and discusses the impact of modern communication technologies on engineering work.

Communication-related topics are discussed throughout this book. Chapter 1 explained the crucial nature of civil engineering work and how it is not enough only to be excellent technically (i.e., civil engineers must be able to explain their work). Chapter 2 outlined the heritage of civil engineering work and explained landmark contributions made by early civil engineers. Imagine the challenge early engineers had in communicating their ideas and persuading the leaders of those times to invest in important civil works, many of which had never been seen before. Chapter 3 explained that the consequences of civil engineering work can be dramatic, with potentially good or bad effects on the public and the environment. For this reason, civil engineers have a special duty to communicate with the public and decision makers about these consequences. Sometimes, as in an environmental impact statement, this involves achieving a balance. For example, is it better to have a secure water supply or to retain habitat?

Chapter 4 described the many types and functions of civil engineering work. Each function (planning, designing, building, coordinating, regulating, and managing) requires communication to ensure success. The design process is central to civil engineering work. Chapter 5 discussed a project of the ASCE, which developed a manual on quality in the constructed project. As a result of the project, the ASCE reported that the single biggest problem in disputes between owners and engineers is poor communication in the process of design and construction. Much civil engineering work is management oriented, requiring effective communication. Chapter 6 outlined management tasks of civil engineers. Management involves many types of communication, from business letters to sophisticated communications of decision information via the Internet and other electronic media.

Chapter 9 will cover government and civic issues, explaining how civil engineers must communicate with the public in many settings to gain approval for projects. Chapter 11 will provide closely related information, such as the civil engineer's role as an expert witness; communication in the legal environment; and the work of civil engineers with attorneys, courts, and various disciplines. Finally, Chapter 12 will examine professional practice and ethics, discussing important aspects of communication, such as issuing public statements in an objective and truthful manner and enhancing the honor, integrity, and dignity of the engineering profession in ways that involve proper communications.

The communications process is a transaction between a sender and a receiver (Fig. 8-1). The sender encodes and sends information, and the receiver decodes and interprets the information, checking the perception of the quality of the communication. The sender determines what is meant and what is to be said, and the receiver must decipher what was

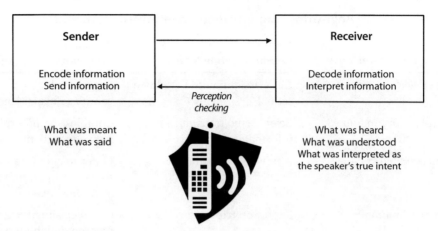

**Figure 8-1**  The communications process.

heard, what was understood, and what he or she interpreted as the speaker's true intent. When this principle is applied to the many venues of communication, the importance of having a clear purpose and of being effective in all aspects of the transaction becomes clear.

It is also important to listen and question well, to clarify, to learn, and to develop relationships. In the management arena, communication is the key to good leadership, and dealing with communication problems is often the most important step in making an organization or relationship the best it can be.

## Venues for Communication

Important communication skills for civil engineers include a number of categories:

- One-on-one or small-group interpersonal discussions (including relationship building)
- Meetings
- Public presentations and reporting
- Internal sales and persuasion meetings
- Brainstorming sessions, including strategizing
- Decision support and advisory sessions
- Coordination discussions
- Writing documents
- Public relations campaigns
- The use of technology to communicate

Table 8-1 shows beginning and advanced communication skills in a number of categories drawn from this list, focusing on written and oral communications.

**Table 8-1** Beginning and advanced communication skills.

| Communication Type | Beginning Communication Skills | Advanced Communication Skills |
| --- | --- | --- |
| Technical report | Class assignment in graduate school | Report for high-level project with complex data and arguments |
| Explanatory memo | Brief memo transmitting information or report | Complex memo explaining new policy to a client group |
| Business letter | Application for first job | Letter to client transmitting sensitive and important information about business |
| Internal memo | Report on trip to see client | Complex situation analysis with recommendations and persuasive arguments |
| Long document | Long paper or report | Textbook |
| Internal presentation | Report to class or in a business meeting | Coordinating an important presentation with multiple parts |
| Speech | Talk before a high school class | Major speech at convention |
| Interpersonal discussion | Discussion with coworkers | Competitive, high-stakes discussion in business environment |
| Sales discussion | Friendly lunch with familiar client | Proposal presentation before critical group |
| Public relations campaign | Resume for first job | Complex public relations campaign to gain funding for major project |
| Use of media | E-mail | Prepare complex web site |
| Meetings | Low-key meeting | Meeting with many agendas and competitive issues |

## Purposes of Communication

Success in any endeavor requires a clear sense of purpose. The first step in communicating technical information is to identify the reason (or purpose) for communicating. There are many purposes behind the need to communicate. If too many purposes are trying to be satisfied at once, a clear focus can become lost and the intended audience confused. When clear goals are established, communication will be more focused and effective. There are many potential reasons to communicate: to inform, explain, persuade, evaluate, problem solve, build relationships, and build teams.

## To Inform

Often, the purpose of communication is simply to provide information. A typical example occurs when an engineering firm acts as a subcontractor to a primary design firm (e.g., a geotechnical consulting firm working for a structural design firm). The geotechnical firm may be determining bearing capacities of a site for a planned structure. Upon completion of the task, the firm is responsible for providing the design information to the structural firm.

The body of such a report is likely to contain numerous tables of data containing results from laboratory experiments. The author's task is to ensure completeness of the information. It is imperative that information that can affect the design is not omitted from the report. The author should report the information clearly and concisely. Readers are likely to miss results that are buried in long sections of prose. Also, readers may not understand results that are vague and open to interpretation. Informative text should be written clearly and concisely and should make use of tabular and graphic presentations.

In the following text, a geotechnical engineering company provides data on soil conditions for a hypothetic building site based on laboratory tests (Table 8-2):

> The average coefficient of consolidation for all soil samples taken from building site 102 was $1.3 \times 10^3$ cm/sec$^2$. These results are the average of four samples tested following ASTM D3245.

This example reports information clearly. The text provides the necessary data and details of test methods, and the accompanying table supports the text.

## To Explain

Civil engineers must often explain a situation or position. For example, a field engineer may decide that construction of a project must be halted until changes are made in the process. The engineer must then communicate to several organizations the purpose for the halt in construction.

**Table 8-2** **Results of consolidation tests.**

| Sample Number | Coefficient of Consolidation (cm/sec$^2$) |
|---|---|
| BH12-3 | $1.6 \times 10^3$ |
| BH12-5 | $1.5 \times 10^3$ |
| BH14-1 | $1.0 \times 10^3$ |
| BH14-2 | $1.1 \times 10^3$ |
| Average: | $1.3 \times 10^3$ |

This communication requires that the position be clearly stated with a supporting rationale. In stating the position, clarity and conciseness are very important. The reader does not want to sift through pages of discussion before reaching the main point, which should be clearly stated at the beginning.

When the reasoning process is presented, the quality of thinking should be reflected in the writing. A steel-trap argument should be presented (i.e., no wiggle room). Writing that rambles and changes directions indicates a poor thought process. It is important to ensure that the presentation is well thought out and follows a logical sequence, as in the following example:

> Construction of the embankment at Site A for the new liquid detention pond has been halted as of 1 January 2000. During routine field inspections, it was found that the required dry density for the embankment of 124 lb/ft$^3$ was not being reached during fill placement. This minimum density is essential to maintain proper permeability of the embankment to meet federal regulations at this site. Placement of fill will not continue until materials and procedures are adopted to reach the required density.

Engineers often serve in the capacity of quality control and therefore must make decisions about construction progress. In this situation, a problem has occurred at a site in which construction methods are not producing the desired results and a decision has been made to halt construction. The reason for the decision (low field measurements of density) is given. Further, the rationale for the decision is strengthened by citation of federal regulations.

## *To Persuade*

Persuasion goes beyond informing and explaining and requires that the reader (or listener) be convinced about something. Usually, this requires starting by either informing or explaining issues, then making a case for the validity of comments. One of the best strategies for persuasive communication is for the writer to take the opposite point of view. To do this, one must first list all the arguments against the thesis from as many perspectives as possible (see Chapter 7). Then each of these issues must be addressed. By anticipating arguments against the thesis first, the points are more likely to be accepted.

The previous example, in which field tests on a construction site resulted in a temporary halt of construction, continues for this example. Here, the engineering consultant attempts to persuade the owners of a potential corrective course of action:

> It is our recommendation to require a new source of fill material before construction of the embankment proceeds. After repeated

field tests found unacceptably low dry densities for the embankment, a series of laboratory tests was performed, including five standard and two modified Proctor tests and three falling head permeability tests. The laboratory test results (see Appendix A) indicate clearly that the current fill material is inadequate for the embankment.

Modified Proctor tests were used to determine whether current fill material could be used with a higher level of compaction energy. However, higher compaction levels in the field will not produce the minimum density. Therefore, it is recommended that a new source of material be required.

The tone in this example has changed from that of the previous example, and there is an attempt to persuade the owner about a specific course of action to rectify a problem. By providing a rationale, the engineer gives a sound technical reason for the new course of action.

## *To Evaluate*

Engineers often evaluate a situation, such as a proposed building site or remediation technology at a hazardous waste site. The communication must convince the audience that the engineer has taken an objective view in conducting a thorough and exhaustive study to determine whether the action being evaluated meets goals. The context in this situation is not unlike that of an adversarial court proceeding: Questions are raised, the standard of proof is given, and the action is weighed against these standards.

# Audiences

Because of the diversity in today's society, civil engineers communicate with many audience types and written and verbal communications must span a wide breadth of styles. The audience's needs, limitations, and vocabulary dictate the style and content of writing. Internal audiences are groups within an organization; external audiences are constituent groups outside an organization.

## *Internal*

Communication within organizations begins with individual communication, such as relationships between peers, workers, supervisors, and different managerial levels. These communications take many forms, including electronic.

Effective engineers, or in fact anyone who wants to succeed in life, will be interested in improving their communication skills. Success in individual communications involves skill. Psychologic techniques (e.g., transactional analysis) have been developed to improve communications, and there are many books on the topic.

Both spoken and written communications must be effective within an organization. Spoken communications often save time and can be interpreted more clearly with the ability to question, interpret, and clarify. Written communications can be saved for reference, shared with others, and amplified to build successively on the original communication.

Communication is important in the functioning of organizations: both in internal work and in relationships with outside constituencies (a critical element of effectiveness in public organization). Effective communication results in coordination of operations. Coordination implies necessity; coordination means to harmonize, and everyone knows that an organization that is not harmonious will not function effectively. Communication and coordination in organizations makes them work better.

Communication within organizations is changing rapidly as a result of computers and communications software (e.g., networking, word processing, e-mail). The ability to communicate better within organizations is leading to some shakeout of middle management. One of the main jobs of middle management is to communicate between strategic levels of management and workers/operators; managers today are able to communicate with more people at a time.

Internal audiences include technical workers and managers, and these two groups require different communication approaches. Language and vocabulary used for management may differ significantly from that used for technical employees. For example, the presentation of weekly progress reports to management will require less technical detail about the design and more emphasis on financial concerns and issues of timing and completion dates. Presentations for management can be challenging to engineers because the language of managers is different from that of engineers. This can make engineers uncomfortable and frustrated. A solution to this problem is to develop a better understanding of the audience's needs and vocabulary. Engineers should take the perspective of managers to be able to communicate better with them. This is a universal rule for good communication: Understand the audience before making a written or oral presentation. Communicating with engineers and technical individuals is a more comfortable situation for engineers because a common vocabulary and understanding are already shared between the groups.

## *External*

One of the most critical and challenging audiences with which civil engineers must communicate effectively is the public. Civil engineering projects usually have a direct impact on the interests of the public, and the public must be kept informed about plans and designs at all stages of a project. The beginning of a project, such as a new highway or water project, will require public support to gain financing. To garner support, engineers must explain the need for the project and its anticipated impact. There are many regulations in place for minimizing the environmental impact of large-scale projects, but the public has a limited understanding of these regulations and engineers must be able to defend the design regarding them.

The use of public involvement in engineering projects requires engineers to listen to the voice of the public. As highlighted in Chapter 3, it is critical that engineers hear what is really desired by the community. Communication is not a one-way street, even in group situations. Ultimately, civil engineers are public servants, whether with government or clients in the private sector. Many voices speak for the public, and hearing them all requires that engineers maintain close contact with such public forums as the local newspaper, television, and radio media. Figures 8-2 and 8-3 illustrate a public meeting in which officials of the Corps of Engineers (Fig. 8-2) are explaining a project to an assembled group (Fig. 8-3).

Communicating risk to the public is also a concern of civil engineers. Risk comes in many shapes and sizes, from contamination of water supplies to an uncertain potential for a devastating earthquake. A common theme is that civil engineers must keep the public properly informed about the risk without causing undue panic or a negative reaction. For example, consider contamination of the water supply. When a water

**Figure 8-2**   A Corps of Engineers presentation at a public meeting. (Courtesy Mobile District Corps of Engineers; photo by Adrien Lamarre)

**Figure 8-3**  Audience at a public meeting. (Courtesy Mobile District Corps of Engineers; photo by Adrien Lamarre)

source becomes contaminated, the public usually must implement such temporary procedures as boiling water before use. This information must be quickly disseminated to all affected populations. Water utilities are required to make public announcements. The media will then seek out civil engineering experts to solicit their opinions and recommendations. Here, engineers must explain complex concepts of risk through an informational style of communication. People need to be reassured during these types of perceived crises, and the style of communication must consider the public's emotions during such incidents.

## Written Communication

The ability of engineers to write well is important early in their careers, and the importance grows as technical and management communications become more complex. Some engineers have become famous for publishing books in areas such as the history of technology.

Engineers will encounter many types of written documents, each with its own particular format and style:

- Letters, memos, and email messages
- Technical and laboratory reports
- Operating procedures
- Policy statements
- Proposals
- Public meeting documents
- Specifications and codes
- Inspection reports
- Graphic presentations

As explained at the beginning of this chapter, engineers should consult helpful guides to learn how to prepare documents, and engineering education should include practice in them. All written documents should be prepared with a few rules in mind: State the purpose, explain the situation, and have a good conclusion. Do not make the documents more complex than they should be, and use language concisely and economically.

## Verbal Communication

Most everyday communication is, of course, verbal. People are said to speak from 10,000 to 25,000 words a day and more. Given that one double-spaced page of written text is about 300 words, then speaking 25,000 words is the equivalent of nearly a 100-page document every day!

The most frequent and important communications are interpersonal. These can range from a simple one-on-one conversation between comfortable colleagues to a conversation with a high-level official. Many techniques can help people speak more effectively. Good verbal communication begins with listening, and there are positive and negative factors to consider. Positive factors include showing interest and understanding, identifying and analyzing the problem from a neutral viewpoint, helping to solve a problem, and avoiding evaluation of the other person's position. Negative factors are illustrated by arguing, interrupting, judging and evaluating, jumping to conclusions, and pushing the other person into a corner.

The average listener understands about 50% of what is said immediately after it is said. Retention drops to 25% within 48 hours and to 10% after one week. Humans can speak at about 125 words per minute but can comprehend speech at about 500 words per minute. These facts help engineers understand why problems sometimes arise in verbal communications and how both oral and written communications must be used to reinforce key points.

Everyone occasionally misspeaks, mishears, and misunderstands. It is important to concentrate on what the speaker is saying without trying to add anything to the meaning. What the speaker has said must then be reflected on. The speaker should be asked for additional clarification when needed. It is important that the person listening understands what has been said and repeats what was understood in his or her own words. It is also important to be aware of cultural influences and differences in communication styles.

Questioning is also an important communication technique. Depending on the phraseology, questions may require yes or no answers or leave subjects open.

Nonverbal communication accounts for much of the total communication. Verbal communication accounts for approximately 7% of what is understood or believed, the way something is said accounts for about 38%, and what the other person sees accounts for another 55%. People stereotype and make up their minds about others within seven seconds of first meeting them.

Providing feedback is an important skill. Never give criticism when angry; instead, allow 24 hours to put things in perspective. Avoid being confrontational or personal, and focus on the true problem. Never take criticism personally, even if it is given personally. Use active listening to try to understand the problem from the manager's point of view. Never immediately agree or disagree with information, and allow 24 hours to honestly assess feedback.

## Public Presentations

Another venue for verbal communication is the public presentation or speech. Public presentations include a wide variety of situations, such as the following:

- Teaching a class
- Speaking before a local board or commission
- Giving an invited speech
- Making a technical presentation at a convention
- Giving testimony to a legislative group

Again, knowing the audience and having a clear purpose will help to plan the presentation. The presentation must be organized with a view to determining its content, duration, main points, and other parameters. Assembling a longer version of the presentation will ensure that the content is adequate; it should then be scaled down and rehearsed. After the length is about right, style, tone, emphasis, and other finer points should be planned. It is important to have effective opening and closing statements, use clear graphics for illustration, pace the presentation and finish on time, engage the audience, and be confident so that any nervousness will not show.

## Failures as a Result of Communication

The importance of communication is evidenced by failures that occur as a result of communication breakdown. For example, messages clearly indicated that an attack on Pearl Harbor on December 7, 1941, was imminent, but no one took action. Another example is the failure of the walkway at the Kansas City Regency. This failed elevated walkway became one of the most serious structural failures in US history.

Communication of design standards in shop drawings was implicated as a principal problem. History is filled with examples of communication breakdowns involving different situations and different forms of communication.

In the case of civil engineering practice, historical failures can sometimes be partially attributed to communication problems. For example, communication failure during construction of the first Quebec Bridge in 1907 resulted in the deaths of about 75 people. The loss of life appears to have been preventable, but a delay in communications may have played a significant role. The story of the failure is presented by H. Petroski (Petroski 1995), who describes that a problem developed during construction of the bridge in 1907. The chief engineer was elderly and ill and was confined to New York. When the first indication of a problem occurred, the field engineer and the chief design engineer corresponded via the mail system of the day. Apparently, this correspondence proceeded for approximately 3 weeks. Eventually, the chief engineer decided that construction should be halted until a resolution to the developing problem was found. Construction continued during this discussion period. On August 29, 1907, a telegram intended to stop construction was either undelivered or unread as the workday ended. Unfortunately, as the workers were finishing that day, the structure collapsed, killing about 75 people.

The loss of life at the Quebec Bridge failure could easily have been prevented using today's communications technology. Imagine that the field engineer had access to a cell phone, fax, or e-mail system. Instead of 3 weeks of letter writing, the communication would have happened much quicker. A digital photograph of the construction site would have been e-mailed to the chief engineer in New York. A discussion of the problem would have taken place over the phone. When the chief engineer realized the seriousness of the situation, the message to halt construction would have been transmitted immediately, with the chief engineer receiving confirmation of receipt of the message via telephone.

Technology and speed do not replace the need to communicate effectively. It is important that the content of communication is not sacrificed for the sake of timeliness.

## Conclusion

Advancing technology has introduced the ability to communicate almost instantaneously with virtually anyone at any time or place. This revolution in communication will impact the practice of engineering in many ways. The quality of communications will continue to be most important, but the way information is delivered is changing rapidly.

Ultimately, the quality of communication is viewed as an indication of the quality of thinking. If the communication is confusing, vague, and not to the point, then the thinking will also be considered confused, unclear, and off-target. The ability to communicate well requires hard work and continual practice. Good engineers continually work on their communication skills.

# 9 Civil Engineers and Government

## Introduction

Chapter 4 presented the idea that the context of the engineer's work is as important as the content. For civil engineers, government often forms the context of and is a key industry in their work. Although not a private industry, government employs many civil engineers as either employees or contractors. Civil engineers also work on projects that are owned or regulated by government.

Without government, society would grind to a halt. Government is an essential ingredient in our lives. Conversely, government can also have a negative influence because taxes are required to sustain it and it sometimes rejects projects. The role of civil engineers as public administrators ideally should be to emphasize the positives and minimize the negatives.

We want to convey the positive side of government work because it is critical to the profession and involves civil engineers so much. Government does not always respond as well as private industry, however, and some negative connotations have arisen. The terms *bureaucrat* and *red tape*, for example, both have negative connotations, although they need not necessarily be negative. The term *bureaucrat* roughly means "one who works in an office" (*bureau* in English or *Büro* in German) but has come to signify a person who adheres too rigidly to rules or regulations. A bureaucracy, by the same token, refers to a government office system with a number of offices and bureaus, each with red tape and formalities.

Red tape was historically used to wrap British government and official papers, and the term has endured to mean any kind of official forms or bureaucratic procedures.

Although we want to show government as a necessary and positive aspect of life, we realize that it sometimes works well only with difficulty. This chapter provides an understanding of how government works (civics) and how civil engineers should relate to government. It also highlights how civil engineers can better work with government and in so doing, make government work better.

## Civil (Public) Works as a Key Area for Civil Engineers

The special responsibility that civil engineers have to work with government stems from the nature of the profession and its close association with public works. Although the dictionary definition of a civil engineer ("an engineer trained to design and construct public works" [*Webster's II* 1984]) is narrow, it does identify the close relationship of civil engineers with public or civil works and hints at how civil engineers must work closely with government and its processes. Civil engineers who gain a good understanding of how government works will find many open doors that can lead to business and career success and will be able to help solve important problems facing the community and nation. Civil works is broader than public works, especially in this era of emphasis on privatization. Government plays a large role in both civil and public works, whether through outright ownership or simply regulation.

Civil works are constructed facilities that are necessary to support civilization, including the built environment, transportation and communications systems, water and energy, and waste management. These broad categories of civil works involve oversight of $10 to $20 trillion in critical assets in the United States alone as well as life-and-death survival issues in developing countries.

Without civil works, civilization could not continue to exist, at least in any advanced form; all levels of government as well as civil engineers must be involved in planning, financing, constructing, operating, regulating, maintaining, and renewing them. Democratic governments must obtain the consent of the citizenry in these activities. Civil facilities do not seem to work as well in nondemocratic types of government. As messy as it is, democracy seems to be the best management framework for public works, as it is for other areas of public affairs. Citizens have strong feelings about their homes, roads, shops, drinking water, and phone systems; and they take an active part in working with the government to manage these facilities. In nondemocratic systems, government may manage civil works without public involvement, but other political problems can emerge.

# Civil Engineers and Civic Engagement

## *Social Capital*

Civic engagement is important to nurture democracy and trust in government in the United States, but some say that the vibrancy of civic life and "social capital" is on the decline (Putnam 1995). Personal responsibility, the indicator of civic spirit that is most closely tied to the state of our nation, is waning (Bok 1996). These opinions are disputed by some, but they do show cause for concern. Civic engagement is everyone's responsibility. Although some disciplines (e.g., political science) are more involved than are others in explaining civic engagement, there are many connections between academic disciplines and civic life. If the disciplines with the most involvement in public life would cooperate in civics education, the impact would be increased.

Civil engineers apply math and science to solve problems. They organize and lead people and control the forces and materials of nature to benefit people. Their overall goal (to advance civilization) is the one that reveals the essence of civic responsibility (ASCE 1998b). The semantics of civil engineering suggest these connections: civil ("of or relating to citizens and their relations to the state"), civil engineering, civil infrastructure systems, civil society, civilized, city, citizen, and citizenship. Areas relating to civics apply, of course, to all citizens. Civics education is a critical component of social studies in the K–12 system.

## *Civics Education*

Americans need to learn much more about civics. Participation in voting and civic affairs has dropped, and such participation is essential for democracy to remain vibrant. According to the Center for Civic Education, about 75% of high school seniors are not proficient in civics. A survey by the LBJ School of Public Affairs at the University of Texas (part of a congressionally mandated national assessment of educational progress) included a question that asked participants to list two ways that a democratic society can benefit from citizen participation. Only nine percent answered correctly. Charles N. Quigley, director of the Center for Civic Education, said this was a consequence of lack of adequate curricular requirements, teacher preparation, and instruction in civics and government *(More Civics, Please* 1999).

In the K–12 curriculum, civics education includes subjects such as civic values, citizenship, government, current political problems, and political problem solving. National education standards for civics and government have developed this list of basic questions to be answered by the K–12 system (Center for Civic Education 1994):

- What is government, and what should it do?
- What are civic life, politics, and government?
- What are the basic values and principles of American democracy?
- What are the foundations of the American political system?
- How does the government (established by the Constitution) embody the purposes, values, and principles of American democracy?
- What is the relationship of the United States with other nations and with world affairs?
- What are the roles of the citizen in American democracy?

Civil engineers must understand these issues because they encounter many civic issues, such as social disintegration, crime and terrorism, economics, environment, race and gender, technology and communication, natural disasters, public health, demographics, and infrastructure deterioration (especially in cities). Although civil engineers are primarily concerned with only some of these issues, the issues themselves are linked to many other issues that do concern civil engineers. For example, the issues of main concern to Americans comprise five categories (Bok 1996):

- Prosperity (economy, scientific research and technology, education, labor market policies)
- Quality of life (living conditions, environment, the arts)
- The chance to achieve (child policies, race, career opportunities)
- Personal security against hazards (crime, health care, regulating the workplace, old age)
- Fundamental values (freedom, personal responsibility, helping the poor)

Other issues with strong and obvious connections to civil engineering include living conditions, the environment, and security against disasters (Bok 1996). As a result of the close relationship between civil engineers and civic engagement, it is important for civil engineers, their partners, and their educators to do more to contribute to civic improvement.

## Civil Engineering Links with Civics

Twelve elements of citizenship can be linked to civil engineering issue areas (Table 9-1). Civil engineers have key roles (1) in explaining how infrastructure and the environment are linked to people's needs and (2) in fostering personal responsibility for involvement of citizens in shared problem solving.

Table 9-1 illustrates that civil engineers can contribute to civics by helping people understand how infrastructure and the environment are important to society's values; helping people understand equity in public services and environmental resources; explaining conservation of resources and security issues related to contamination, sabotage, and disas-

**Table 9-1    Citizenship requirements and responsibilities.**

| Citizenship Requirements | Citizen Responsibilities for Infrastructure and Environment |
|---|---|
| Understand society's values | Understand links between civil infrastructure systems (CISs), environment, economy, and social welfare |
| Know about country, government, society | Understand how CISs and environmental systems work |
| Understand the rights of all | Understand rights to public services and a clean environment |
| Understand personal responsibility in society | Conserve resources/prevent contamination/sabotage/disaster |
| Defend justice | Provide public services, environmental equity, public safety for all |
| Participate in democracy to solve society's problems | Get involved publicly in selecting solutions for common good |
| Be fair to all, sharing, and interacting with others | Share space and facilities and extend public services to all |
| Know about society's problems | Support CIS and know impacts on society/environment |
| Pay taxes and fees | Pay fair taxes and fees |
| Cooperate with government and institutions | Be willing to renew and replace CIS and environmental systems |
| Provide mutual assistance in community | Help others solve CIS problems and share in management |
| Support civic institutions | Have civic spirit in cleanups, promotion of public facilities |

ters; organizing public involvement; identifying shared solutions to public works and environmental problems; explaining the need for funding of civil infrastructure systems (CISs) and environmental protection; explaining why CISs is needed; and participating in community programs. Table 9-2 illustrates how civil engineers can contribute to civics education.

## Structure, Functions, and Organization of Government Agencies

### Structure

Although students are taught about government as they progress in school, most adults have a limited view of government. By contrast, civil

**Table 9-2  Civil engineering contributions to civics education.**

| Areas | Players | Activities | Results |
|---|---|---|---|
| Broad civics education | Team of consultants and public works officials | Explain to citizens and students how infrastructure improves economy and environment | Citizens and students gain insight into complex interactions between their actions and consumption and taxes and environmental impacts |
| Social and environmental justice | Educators and community groups | Work together to study social and environmental equity issues | Results reported to public agencies and regulator groups leading to better understanding of need for action |
| Conservation and security | Consultant, regulator | Consultant explains links between flood project and security of low-income residents; regulator explains impacts of conservation on lowering taxes and fees | City decision makers understand the need for actions to improve floodways and to take conservation seriously because of the fiscal benefits |
| Public involvement in problem solving | Consultant, public official | Work in cooperation to bring latest public involvement techniques to draw in citizens and students to make infrastructure decisions | Citizens and students are more engaged in civic life, and confidence is raised in results |
| Sense of community | Public officials | Cooperate to link systems to increase security (e.g., against computer failures) | More confidence in public services, appreciation that public officials can work together on shared problems |
| Citizen trust in government | Consultant and public official | Involve citizen groups to develop better explanation of taxes and fees | Improves citizen trust, leads to more value in taxes and fees, and improves management in public works agency |
| Public spirit | Consulting firm | Takes lead in organizing student groups to work with service club on clean up of river banks | Involves engineers with environmental groups and students, leading to better relationship and appreciation of what engineers do |

engineers require far more than average knowledge about government. It is important to understand government structure (the form government takes and the processes it follows). Government is complex, including many functions, levels, locations, and other characteristics. The United States is a republic (i.e., a nation of states held together by a central government [the federal government]) and a representative democracy (i.e., the people rule through elected representatives).

**Table 9-3**  Roles in governmental branches.

|  | Executive | Legislative | Judicial |
|---|---|---|---|
| Federal | President | Congressman | Federal judge |
| State | Governor | Legislator | State judge |
| Local | Mayor | City council member | Local magistrate |

Government is comprehensive and involves the processes, organizations, powers, and institutions that establish laws and policies and place them into effect so that society can function. It provides for rule of law, processes of government to take place, certain services needed by citizens, and order keeping. Modern society often takes the rule of law for granted; however, even today barbaric acts still occur when governments break down. This can occur in societies in which despotic rulers are in charge of government or in societies where there is civil unrest.

The US government has three branches (executive, legislative, and judicial), all of which exist at the national (federal), state, and local levels (Table 9-3). Civil engineers are involved at each level with each branch, although most of their involvement is with agencies of the executive branch at the local level. An example of this is employment with the public works or planning department of a city government in which approval is required for a civil engineering project.

The US democratic system of representative government is important to civil engineers because they work directly with the people in programs of public involvement. Civil engineers also rely on representatives (e.g., legislators) and congressmen to make decisions. The work of public agencies is largely carried out by officials and workers who have been selected through a civil service or merit system. Elected officials also participate, as do appointed officials who are part of the political component of government as opposed to career civil service workers or bureaucrats. Civil engineers work with three categories of public officials: elected, appointed via the political process, and appointed via a merit system. These categories of public officials respond to different incentive systems, and it is important to understand how they got their jobs and what they must do to keep them or to advance.

## *Functions*

Civil engineers work more than do other types of engineers with the processes and institutions of government and therefore must have a good knowledge about political science. Civil engineering students, however, rarely take courses in political science during their undergraduate studies. The basic concepts of political science (the study of the principles, processes, and structure of government and political institutions) include voting; the role of citizens; legislative process; and roles of public officials such as mayors, police officers, and judges.

Voting is an important right in the United States. In addition to voting for representatives, citizens increasingly have the right to vote on specific civil works projects. This can be important for civil engineers if, for example, the client must gain voter approval for a project that is being managed by the engineer. Civil engineers can find themselves in a political campaign in order to gain approval of a project. With the trend toward direct democracy, civil engineers should expect to work more, not less, with government in the future.

## Organization of Government Agencies

The structure of government is revealed in the names and functions of its agencies. In the executive branch, relationships can be traced among similar agencies across the three levels (Table 9-4). For example, in a transportation issue, the agencies involved are the Federal Highway Administration, the state Department of Transportation (DOT), and the city transportation department. The other examples in Table 9-4 follow to some extent, although it is not always a straightforward issue (e.g., buildings). Federal agencies do not always get involved; however, most regulatory activity occurs at the federal level.

## Federal Government

At the federal level, there are many government agencies, but the main cabinet level agencies are the following (United States Federal Government Agencies Directory 1999):

- Department of Agriculture
- Department of Commerce
- Department of Defense
- Department of Education
- Department of Energy

**Table 9-4** Examples of relationships between governmental agencies.

| Transportation Issues | Water Issues | Building Issues | Energy Issues |
|---|---|---|---|
| Federal Highway Administration | Corps of Engineers | National building code | Federal Energy Regulatory Commission |
| State department of transportation | State department of water resources | State building code | State public utilities commission |
| City transportation department | City water department | City building department | Local electricity department |

- Department of Health and Human Services
- HUD
- Department of the Interior
- Department of Justice
- Department of Labor
- Department of State
- DOT
- Department of the Treasury
- Department of Veterans Affairs

In addition, a variety of independent agencies are of special interest to civil engineers, including the following:

- EPA
- FEMA
- General Services Administration
- National Aeronautics and Space Administration
- NSF
- National Transportation Safety Board
- Nuclear Regulatory Commission
- Occupational Safety and Health Review Commission
- Panama Canal Commission
- Peace Corps
- TVA
- Agency for International Development

Government uses the political process to debate its priorities in providing services and to budget for government expenditures. In theory all these agencies report to the president, but in practice agencies respond to congressional oversight committees, the general control mechanisms of government (e.g., Office of Management and Budget), and White House staff. The federal government is simply too large and complex for all parts to receive the personal attention of the president.

## State Government

State government agencies mirror those of the federal government, with some exceptions. States have different names for their agencies, although the functions are similar. Table 9-5 presents a comparison of the main state agencies in Colorado and North Carolina (not all functions are shown).

Civil engineers work with state agencies in much the same way they do with federal agencies. Civil engineers may also encounter state-to-state or interstate issues in their work. State governments are becoming more important as the complexity of government increases. They form a buffer between the federal and local levels and get involved in almost

**Table 9-5**  Comparison of some Colorado and North Carolina state agencies.

| Colorado State Government | North Carolina State Government |
|---|---|
| Agriculture | Agriculture |
| Corrections | Correction |
| Education | Public Instruction |
| Human Services | Health and Human Services |
| Natural Resources | Environment and Natural Resources |
| Personnel/General Support Services | Administration |
| Public Health and Environment | Health and Human Services |
| Public Safety (and State Patrol) | Crime Control and Public Safety |
| Revenue | Revenue |
| Transportation | Transportation |

(Compiled from the State of North Carolina web site: http://www.state.nc.us; accessed 3/15/99)

every issue that the federal government does. You can learn something about state government by viewing programs of the Council of State Governments.

Most of these issues impact engineering, particularly natural resources, environment, facilities, and transportation.

## *Local Government*

The functions of local government agencies are oriented to meeting local needs such as infrastructure, police, fire, and local services. There are some 85,000 local government entities in the United States, including municipalities, county governments, special purpose districts, regional governments, and other entities. Civil engineers do a lot of business with local governments and can profit from understanding them. The western city of Fort Collins, Colorado, for example, has a city manager form of government with the following major departments (City of Fort Collins, Colorado 1999):

- Arts and Culture
- Citizen Services
- City Hall
- Community Planning
- Environment
- Library
- Parks and Recreation
- Safety

♦ Transportation
♦ Utilities

In the local government of Fort Collins, engineering work falls within the Transportation and Utilities departments. Transportation includes the bus system, traffic control, planning, and the Engineering Department:

> [The Engineering Department] provides a variety of services relating to the construction and design of the City's capital projects and subdivision development. Among the responsibilities of Engineering is the provision of construction management which includes planning, design, and inspection services for capital projects constructed in the public right-of-way. It manages street maintenance (overlay) through the Pavement Management Program. The department also provides surveying, drafting and mapping services for other City departments. Engineering also has the responsibility for management and coordination of subdivision activities, including the creation of Special Improvement Districts (SIDS) and Street Oversizing (City of Fort Collins, Colorado 1999).

Although in this example the Engineering Department falls within the Transportation Department, engineering activities include functions that go well beyond transportation.

The Fort Collins Environment Department includes air quality, natural resources, natural areas, xeriscaping, recycling, solid waste, wind power, and environmental education. Its web site announces that "the City has many programs that enhance and protect the land, the water, and the air, from pollution prevention to wildlife habitat enhancement" (City of Fort Collins, Colorado 1999). A taxpayer might ask if these services are essential; however, it must be remembered that the department is small and the quality of life in the city is consistently rated highly.

The city of Charlotte, North Carolina, a large, eastern city, has a different form of government. It cooperates with the County of Mecklenburg and has major departments (some are joint city and county departments) with the following names (City of Charlotte, North Carolina 1999):

♦ Economic Development
♦ Engineering and Property Management
♦ Finance
♦ Fire Department
♦ Human Resources
♦ Information Technology
♦ Land Development
♦ Neighborhood Development
♦ Parks and Recreation (city/county consolidated)
♦ Planning (city/county consolidated)

- Police (city/county consolidated)
- Public Service and Information
- Purchasing (city/county consolidated)
- Solid Waste Management Services
- Storm Water Services
- Transportation
- Utilities (Charlotte–Mecklenburg Utility District) (city/county consolidated)

San Francisco is an older, West Coast city, and its departments include the following (City of San Francisco, California 1999):

- Airport
- Building Inspection
- Embarcadero Project
- Department of the Environment
- Hazardous Waste Management Program
- Housing Authority
- Municipal Railway
- Parking and Traffic
- Planning
- Police Department
- Port of San Francisco
- Public Utilities Commission
- Public Works
- Recreation and Parks
- Redevelopment Agency
- Transportation Authority

Notice that San Francisco retains the department title "public works," whereas the other cities no longer use the title.

## An Example of How Government Works

To explain how government works in ways that are meaningful to civil engineers and their clients, consider a hypothetic project (a new subdivision) and identify some of the ways government gets involved.

First, the subdivision must be conceived by someone, probably the developer. To check property availability, the developer studies land transactions at the county courthouse (local county government), where the real estate office or tax assessor maintains the records. If the land is located inside the city limits, the developer will inquire with city planners and engineers (local city government) about zoning and other rules. They, or the city finance office, should inform the developer about required fees and any capital projects pending. If intercity transportation projects are involved, the state DOT (state government) will be involved.

If water rights, as administered in Colorado, are involved, the State Engineer's Office (state government) will have the information. Perhaps a stream goes through the property and will require remedial work that involves a dredge and fill permit, administered by the Corps of Engineers (federal government). If wetlands are involved, this could involve the EPA (federal government) or the Corps as well. If endangered species are mentioned, the Fish and Wildlife Service might require consultation (federal government).

These are examples of executive branch agencies at the three levels of government. The types of agencies mentioned are tax assessor, planning and zoning, engineering, finance, transportation, state engineer, Corps, EPA, and environmental (Fish and Wildlife). Legislative branches of government for the three levels might also be involved, but this is not as common. Civil engineers might be involved, for example, in briefing elected city council members about a potential project. Unless they approve, the project might not proceed. A state legislator might be asked to get involved with a state agency that was involved in a grant application or some kind of approval, and in large developments congressional issues could conceivably be involved, although this is not common. Hopefully, judicial branch agencies will not be involved in this example, but federal and state courts, municipal courts, and hearing officers for appeals of agency decisions are sometimes involved. Civil engineers will not work as much with the judicial branch but might become involved in litigation, usually representing one side or another in an action such as an appeal of a permit denial. Expert witness work involves civil engineers, testifying in civil cases in which one side is suing another about a project or damages.

Many government offices and functions are involved in civil engineering projects, and engineers should understand the reasons for each government function, how it affects the project, and how to make it work better.

# Planning, Coordinating, and Decision Making

Civil engineers work especially closely with government in planning, coordinating, and decision making. This includes activities involved in deciding what to do, involving the stakeholders, and getting permission (e.g., in civil engineering projects).

Planning is a very complex activity involving many players, decisions, and outcomes. For a simple project, such as replacement of a small bridge or expansion of a small water treatment plant, project planning must fit into existing comprehensive and capital plans and conform to constraints imposed by the client and regulatory agencies. Large projects may require extensive planning processes in which major decisions affecting millions of people are required. These projects can go on for years and involve major political issues and actions.

Regardless of whether the planning process is simple or complex, government will control it in one way or another. To illustrate, consider three types of plans that civil engineers encounter often: comprehensive development plans, master plans for infrastructure facilities, and plans for specific projects.

## Examples of Plans

Comprehensive development plans seek to integrate the many issues involved in regional development, including land use, natural areas, transportation, water and wastewater systems, energy supply, waste disposal, communications, cultural facilities, recreation, and other societal considerations. The field of urban and regional planning exists to advance these types of plans, and there is an extensive body of literature about how to approach them. Government is deeply involved in comprehensive development plans. Local government is responsible for coordinating the plans, state government will relate to them through transportation corridors and other matters of state interest, and the federal government will have an interest because of extensive regulatory mechanisms.

For the same reasons, master plans for infrastructure facilities also involve all three levels of government. These plans are to guide the development of transportation, water, flood control, and other facilities to support comprehensive development plans. An example might be a water supply master plan for a community. A specific project plan might be derived from the master infrastructure facility plan. Taking the water supply example further, a new treatment plant might be planned through the conceptual and preliminary phases, before drawings are prepared.

## Coordination as a Function of Planning

Coordinating is one of the reasons for the existence of government. It is the primary way to involve the project stakeholders and achieve harmony among them. The definition of *coordination* is achieving harmony among diverse interests. Civil engineers will find that learning to be an effective coordinator among many competing interests is an important ingredient for success. Elements of negotiation and mediation can be important parts of this sensitive and important process.

Communication is a main ingredient of coordination, involving meetings, personal contacts, discussions, and other avenues. The Internet has become a very important tool for increasing communication. In planning, decision making is a main activity of government, especially the executive branch. Executives in the public sector make many decisions every day. Decision making is covered more fully in Chapter 5.

# Evolution of Government Roles

The evolution of the US infrastructure reflects the phases of our national development: the early agricultural society, the emerging nation, the Civil War, the Industrial Revolution, the advent of the automobile, the Great Depression, World War II and the postwar urban explosion, the suburbanization and road building era, and the shift to the Information Age. Each phase involved many political questions related to infrastructure.

For example, whether the federal government should finance internal improvements (e.g., roads and canals) was a question that divided the nation from its earliest days. Sectionalism and varying tax policies related to internal improvements, or public works, contributed to the onset of the Civil War. During and after the Industrial Revolution the nation grappled with the question of public versus private power (privatization).

As a nation develops, the balance of power may shift from national to local government and vice versa. In the early days of the United States, state governments were somewhat autonomous and state–federal relationships (including states' rights) were key issues in the Federalist debates. Today, that battle continues, with many issues having been settled and new ones arising regularly. In developing countries it is normal to find a strong central government and weak local governments. Building the capacity of these local governments is a key goal of development strategies.

Participation of federal, state, and local governments in the US economy can be measured by their budgets. Each level of government handles unique expenditures. For example, the federal government pays for defense, interest on the national debt, welfare, pensions, and many agency programs. State government also includes welfare, plus education, health and hospitals, and highways. Local government pays for police, fire, community facilities, public works, and myriad other functional units.

In 1995, revenues for the three levels of government were as follows (US Department of Commerce 1998):

- Federal   $1,573 billion
- State     $906 billion
- Local     $757 billion

Local government is most dispersed, with some 87,453 units in 1997 compared with 50 units for state government. Expenditures for public works in 1995, including highways, airports, water transportation, sewerage, solid waste, water supply, and mass transit, were as follows:

- Federal   $12.8 billion
- State     $56.4 billion
- Local     $123 billion

The role of government is debated continually, and the possibility of privatization continues to be discussed. One of the eye-opening aspects of privatization has been the experience of Russia and Eastern Europe after the death of communism. Markets work, but privatization is not easy. Civil society is seen to be crucial, with a real need for a vibrant third sector. The third sector includes nonprofit organizations (e.g., associations of professionals, churches, chambers of commerce, charities). Geography and history influence the pace and depth of transformation, and initial conditions matter (often in unexpected ways). For Eastern European nations, admission to the European Union has created new demands for infrastructure, including highways, power lines, communication links, and fiber-optics.

## Congressional Committees

The apex of policy work is in Washington, of course. Policy issues are discussed there with congressional committees, their staffs, and related professional groups. Conventional civics explains how bills become law, but Congress does much of its work in committee. Many committees affect civil engineering in some way. Most committees are for authorizing legislation or for oversight of federal activities. Appropriation committees deal with actual financial operations. Examples of congressional committees that influence civil engineering work are shown in Table 9-6.

Civil engineers involved in policy work may be in contact with these committees or their staffs when visiting congressional offices in Washington (i.e., Capitol Hill). The work of congressional committees related to civil engineering can be viewed through the evolution of the House Transportation and Infrastructure Committee or the Building Committee of Congress (US House of Representatives Committee on Transportation and Infrastructure 1999).

## Policy Issues

Issues such as these are dealt with in what is known as the policy process (the process of making rules and setting the framework for action). Policy clusters, or networks, are products of government action and refer to groups of influential people interested in similar topics such as highway finance (Hofferbert and Cingranelli 1996.) Civil engineers can choose to get involved in government through the policy process. Various levels of policy can be influenced to promote improved facilities and environmental control. Policy exists at all three levels of government.

**Table 9-6** Congressional committees that influence civil engineering work.

| Civil Engineering Areas | House Committees | Senate Committees |
|---|---|---|
| Soil conservation | Agriculture | Agriculture, Nutrition, and Forestry |
| Project funding | Appropriations | Appropriations |
| Corps of Engineers | Armed Services | Armed Services |
| Urban development | Banking, Finance, Urban Affairs | Banking, Housing, Urban Affairs |
| Capital budget issues | Budget | Budget |
| Energy development | Energy and Commerce | Commerce, Science, Transportation |
| Natural resources management | Natural Resources | Energy and Natural Resources |
| Transportation (House side) and Environment (Senate side) | Transportation and Infrastructure | Environment and Public Works |

The ASCE has an active Washington Office that tracks policy issues. The ASCE web site, **www.asce.org** lists 155 policy statements. These categories illustrate the range of policy issues considered important by leading civil engineers (ASCE 1999):

- Coastal zones and waterways (e.g., beach erosion)
- Disaster mitigation and response (e.g., earthquake engineering research funding)
- Education (e.g., first professional degree)
- Energy (e.g., hydropower licensing)
- Engineering practice (e.g., collective bargaining)
- Environmental issues (e.g., Endangered Species Act reauthorization)
- Government (e.g., professional grade salary structure for government engineers)
- Hazardous waste (e.g., Superfund reauthorization)
- Infrastructure (e.g., infrastructure impacts of population growth)
- Legal reform (e.g., professional liability and tort reform)
- Procurement and contract issues (e.g., qualifications-based selection of engineers)
- Public involvement (e.g., appointment of engineers to policy positions in government)
- Public works financing (e.g., Davis-Bacon Act)
- Quality and standards (e.g., uniformity of building codes)
- Registration (e.g., professional licensure mobility agreements)
- Research (e.g., federal incentives for construction innovations)
- Safety (e.g., responsibility for dam safety)
- Social issues (e.g., minority- and women-owned businesses)
- Solid waste (e.g., source reduction of solid waste)
- Space (e.g., engineering and construction in space)
- Surface transportation issues (e.g., intelligent transportation)
- Tax issues (e.g., taxation of employee-paid educational assistance)

- Waste water (e.g., Clean Water Act reauthorization)
- Water resources management (e.g., emergency planning by water providers)

# Regulation of Projects

Recently, there has been a shift in the work of civil engineers from emphasizing only building projects to embracing the more complex role that includes additional regulation. To regulate means to control. Regulation refers to government as a mechanism to coordinate behavior and ensure the rights of individuals. Society is much more regulated today than in the past. Regulation involves everything from speed limits to gun control to limits on the number of fish that can be caught in a day. Regulation in civil engineering involves aspects of projects, land use, water controls, and many other issues. Projects are regulated by all three levels of government, and most regulations are in the categories of land use, environmental, health, and financial limitations (Table 9-7).

Much of the work of civil engineers is involved with regulatory controls. A few specific examples illustrate this:

- In seeking to build a new water supply reservoir, a community may have to apply for a federal 404 permit. Opposition forces will use the permit process as a way to defeat the project or impose conditions. The civil engineer will prepare maps, studies, evaluation material, and permit applications to assist the client.
- In developing land, an owner may need to mitigate potential threats to wildlife under the Endangered Species Act. The civil engineer might work with other disciplines (e.g., wildlife specialists) to study habitat and wildlife biology so that a mitigation plan can be developed along with the plans for developing the land.
- Under rules of the Safe Drinking Water Act, a city may have to modernize a water treatment plant. The civil engineer might have to determine what features in the new plant will satisfy the rules.
- In designing a new subdivision, a civil engineer might have to work with the city engineer's office to set up street sizes and designs that meet the city code.
- Open space requirements, established in a city code, might dictate the work of a civil engineer in laying out the details of a new subdivision and influence everything from street alignment to the number of lots that can be provided.
- Minimum pipe sizes are normally established in codes and standards of various kinds, illustrating regulatory controls that link government rules to standards established by professional groups, such as the ASCE. A civil engineer will encounter these, for example, when providing water supply pipes that meet fire codes.

**Table 9-7**  Categories of governmental regulations.

| Category | Examples Of Projects |
|---|---|
| Land use | Number of lots per acre in a subdivision |
| Environment | Water quality rules |
| Health | Clean air limits |
| Finance | State regulatory commission controlling electric rates |

- Building controls will govern issues such as drainage away from a structure, height of a building, and many internal design issues that involve architects as well as engineers.
- Storm water runoff controls have increased greatly in recent years, mainly to respond to Clean Water Act rules but also to respond to flooding and open-space controls.
- Impact fees illustrate a financial area of control that civil engineers will encounter as they seek to provide cost-effective development strategies. These are usually linked to growth-pays-its-own-way philosophies of regulatory controls.

Civil engineers will encounter many other examples of regulatory situations, mostly stemming from laws passed to respond to public pressures for solving various societal problems such as water pollution, urban sprawl, air pollution, solid waste disposal, and cost of living.

## Financing

Government is also involved in project financing, especially for large and public projects. Examples of ways that government is involved include the following:

- Government budgeting
- Issuing bonds
- Public utility controls
- State infrastructure banks
- Taxes and fees
- Government appropriations
- Government accounting
- Financial policy
- Privatization

## Public Involvement

Civil engineers have a great opportunity to help achieve public involvement goals through work to foster public involvement in the planning of effective projects. Civil engineers are key professionals to lead public participation programs. Public involvement begins with individual citizens, making sure that each citizen knows what is planned and has a say in the planning process. This might begin with neighborhood meetings and proceed to more formal presentations before elected boards and commissions.

Although public involvement can involve complex issues, it is a simple process to include the public in planning and decision making. The challenge of making democracy work is the involvement of the people in decisions that impact them. Today, there is much discussion about people being disenfranchised from public decisions. At its best this simply means that people do not participate; at its worst, it might mean they actively oppose legitimate government.

One part of the complexity lies in defining who comprises "the public." It is not enough for the public to consist only of influential citizens who have a financial stake in decisions. Nor is it enough for the public to consist only of representatives of interest groups that are loud and visible. The public should be defined to include all stakeholder groups. They might be stakeholders by virtue of proximity, economic stake, use, social concern, or value systems.

There are numerous guides (primarily government manuals and journal articles) about how to achieve public involvement. One such example is a manual by James L. Creighton, who wrote from the viewpoint of someone who had been "bloodied" in the process. In his book, Creighton refers to the fact that he has been in many divisive public meetings. He traces the origins of government-required public involvement to the War on Poverty in the 1960s and the environmental legislation of the 1970s. As a result of these acts and others, public involvement requirements became institutionalized, and are now widely accepted. Of course, it always made sense as a practice of democracy to involve people in decisions that affect them. That is the idea behind a representative democracy. Creighton defined public involvement, whatever its name, as "a process, or processes, by which interested and affected individuals, organizations, agencies and government entities are consulted and included in the decision-making of a government agency or corporate entity" (Creighton 1981). Technical means by which public involvement can be accomplished include individual interviews, workshops, advisory committees, informational brochures, surveys, and public hearings (Creighton 1981).

Engineers must watch out for their sense of the public being in the way—on the contrary, the public is the customer. It is easy to get carried away with designing "optimal solutions" that do not make sense to the public. It is important to present people with information and alternatives in language they can understand and further to involve them sincerely in improving the understanding of problems, defining solutions, and making decisions.

That does not mean that everyone must agree with everything—they never will. It means to involve the public and then to make the best decision considering all points of view. One way for engineers to involve the public is to volunteer in public schools (Fig. 9-1).

## Politics

In working with government, civil engineers will quickly become aware of politics. There are many reasons for politics, of course, and civil engineering projects often involve convergence of powerful political forces. Politics has several shades of meaning, from the science of government to using intrigue to maneuver within a group.

When does government end and politics begin? Given that the purposes of government are to keep order and provide for the rule of law, processes of government, and services needed by citizens, it is natural that political issues will intrude into the arenas where civil engineers work.

One way to view politics is as a group of agendas—of individuals, interest groups, political parties, and stakeholders in general. Individuals and groups seek what they consider as favorable outcomes in public decisions such as where road improvement funds are spent, whether a water project is approved, whether taxes are raised, or even whether they are elected to office. The individuals or groups then use influence to improve their chances of gaining the outcomes they seek.

Some of the issues civil engineers must consider include the influence of political campaigns on projects, motivations of elected officials, roles of lobbyists and interest groups, and use of associations to influence policy. The example of how public officials get their jobs (via elections, political appointments, or civil service) illustrates the motivational factor.

Although everyone has a civic duty to be involved in the political process, engineers normally do not get involved in political campaigns in conjunction with their engineering work. Engineer involvement in political campaigns such as these would be in his or her role as a private citizen, not in the office of engineer. There might be situations, however, in which an engineer gave presentations to groups to promote a bond issue, which some groups oppose politically. In situations such as this, the engineer should ensure that he or she is complying with the code of ethics and that all conflicts of interest are disclosed. Running for office is a fine way to exercise citizenship, and there is no reason that engineers should not use their skills (e.g., clear thinking and effective decision making) as a way to serve the public in elected office.

**184** ♦ *Civil Engineering Practice in the Twenty-First Century*

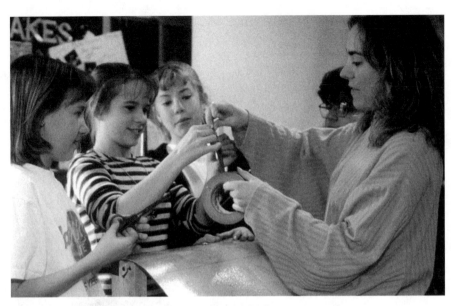

**Figure 9-1** Engineer explaining a project to K–12 students.

Perhaps the politics faced most often by civil engineers are interest group politics, in which groups promote or oppose a particular action being worked on by the engineers. Growth and environment issues that civil engineers commonly face fall into this category. Typical issues include NIMBY opposition to any plan and no growth issues. Regional competition is another example. The situation may even become one of ethics: Does the civil engineer promote the project and one or more interest groups, or does he or she remain completely objective and represent all sides?

At a much more complex level, the *triangle* refers to interest group phenomena such as what President Eisenhower called the military–industrial complex, which consisted of military officers wanting weapons, defense contractors wanting business, and politicians wanting influence or to push their agendas. Eisenhower warned in the 1950s that this military–industrial complex might gain too much influence in the United States and even carry us to the brink of war. This phenomenon exists in all policy areas in which an interest group gets to know part of the bureaucracy that depends for its jobs and power on the issue area being important. It then applies pressure to a legislator, who gets attention for being a champion in this area.

There are other places in which politics enters life as well, such as in bureaucratic, organizational, or office maneuvering; but these areas are beyond the scope of this chapter. What we might say is simply that relationships and lining up support for your ideas and points of view are important in everything you do. Sometimes people define this as politics;

it is, however, not negative as long as everything is ethical, above board, and free of conflicts of interest.

Government decision making is often characterized as incremental because each step is built on the previous one as opposed to being a dramatic change in direction. Incrementalism sometimes frustrates engineers who like to set goals and move directly toward comprehensive solutions and reforms. People who are politically aware (or think they are) can sometimes be contemptuous of engineers who (they believe) are naïve and only cognizant of technical topics.

Budget politics are of concern to managers at all levels of government, especially those involved in infrastructure, where so much money is at stake. In any organization the budget becomes the focal point where agendas are worked out with internal competition. For the US budget, this competition takes on global dimensions. Budget politics involves the triangle of interest groups consisting of agencies, politicians, and public interest groups. The interest groups work through the agencies and/or politicians to try to get budget resources invested in their favorite areas. At a simple level, a group wanting a street paved (interest group) may see the city engineer (agency), who may want the street paved also but lack the budget, and see their elected representative, who may advocate putting that item in the city capital budget. Through this triangle of interests, the project may get done.

Political issues in budgeting start with the agency's roles and expectations and in deciding how much to ask for and how much to spend. Then there is competition within agencies for permission to request budget amounts (departments versus bureaus). The budget office has a strong role in deciding how much to recommend, and in the legislatures the appropriations committees have roles and perspectives in deciding how much to give and how to respond to client groups. Client groups must be dealt with in a comprehensive manner. They range from the elderly with an interest in social security to environmentalists who lobby for wastewater plants. At the local level, developers watch the budget process to see what they must pay for infrastructure. The antigrowth forces want these fees to be high.

Civil engineers cannot completely avoid all politics, of course, such as in participating in government processes. Some aspects of politics must be avoided, however, such as seeking work in improper or unethical ways. Chapter 12 addresses these issues in more detail. Engineers must realize that the political process has little to do with expertise (the main quality of engineers) and much more to do with relationships, political motivation, coalitions, and other nonengineering subjects. Engineers can be involved in politics, but they do not inherit this right by being engineers.

# 10 Economics and Finance for Civil Engineers

## Introduction

Civil engineers apply finance and economics principles in many ways to life and to professional practice. Both finance and economics use monetary values: Finance deals with how to pay for things, and economics deals with decisions about allocating society's resources. In addition, economics is a tool for analysis, forecasting, education, and decision making, whereas finance has more focused management uses.

Economics deals with many subjects and gives rise to good-natured jokes about the inability of economists to reach conclusions. The reason for this is that economics must take into consideration so many societal trade-offs, such as poverty and wealth, national unemployment, and the rate of growth of states and regions. That is the nature of the field. Civil engineers use economics a great deal, and it can be an extremely useful subject for study. Finance involves practical decisions based on the bottom line (e.g., profit or loss, rate of return). As civil engineers advance in their careers, particularly if they become managers, finance becomes a very practical subject. Skills such as preparing a budget, seeking revenues, controlling costs, and reading financial statements are essential.

## Economics for Civil Engineers

Civil engineers need economics mainly to understand the society in which they work. They use tools available from this key discipline to be more effective citizens and to serve the public better through professional practice. This chapter presents key concepts of economics simply and briefly and outlines the aspects of economics that affect civil engineers most directly.

Economics is an important subject for civil engineers to understand. In addition to explaining issues that are important to all citizens (such as taxation and government expenditures), economics explains two problems of fundamental interest to civil engineers: the link between infrastructure and the economy and how to make decisions about the protection of the environment. Economics is the art and science of deciding how to allocate society's resources to meet its demands. Economics also helps society find a balance in the use of land, labor, structures, vehicles, equipment, and other resources, including the environment. Economics provides an understanding of how society makes resource allocation decisions.

Using the work of economics professors Robert Heilbroner and Lester Thurow as a guide, a road map to a basic understanding of economics can be created (Heilbroner and Thurow 1994). Elements include the following:

- Capitalism versus socialism
- Famous economists (e.g., Adam Smith, Karl Marx, John Maynard Keynes)
- How the economy works
- Government in the economy as regulator
- Government in the economy as service provider
- How economic trends are measured
- Gross national product (GNP) and economic accounting
- Savings, investment, and consumption
- Public sector economics
- Money and banking
- Inflation
- Employment
- Stimulating the economy and labor economics
- International economic issues
- Central issues of economics for civil engineers
- Productivity
- Markets
- Poverty and wealth
- International economics
- Depressions and recoveries

## *Capitalism and Socialism*

Behind the economic issues of concern to civil engineers is the basic difference between socialism and capitalism. The United States has a mixed system of public sector (socialistic) and market (capitalistic) economics.

Because civil engineers in the United States work on problems of the government sector, their work is often indistinguishable from that in more socialistic economies. In other words, civil engineers work largely on problems involving public sector economics rather than market economics. For example, in work on a large bridge owned by a government authority (e.g., the Port Authority of New York and New Jersey), both the client and the regulatory rulemaker would be government but the engineers and contractors would be from the private sector—a mixed approach.

Currently, there is a trend toward privatization of public services, shifting the pendulum more toward the capitalistic viewpoint. Even this does not completely place things in the market environment because some services (such as water supply) are primarily monopolies, and the main difference between a public monopoly and a private monopoly is how the service is regulated. Economists such as Smith, Marx, and Keynes form "book ends" of eras when different economic systems were attempted. These economists debated the merits of capitalism and whether self-interest or state intervention works better to meet society's needs. The main argument for capitalism is that self-interest will meet the needs of society and of individuals.

Marx advocated using socialism to reform the abuses of capitalism and became the intellectual guru for the move toward global communism. His story is a long one of ideologic struggles to implant the socialistic point of view, or anticapitalism. Keynes offered a way to use government intervention in the economy to regulate and smooth out the drastic cycles of market capitalism that were experienced again and again in depressions and downturns in the United States and throughout the world, especially the Great Depression of the 1930s. Keynes' ideas formed much of the intellectual capital of the New Deal and stimulated much construction and related civil engineering work. The Great Depression led directly to public works agencies such as the Public Works Administration and Works Progress Administration. During the 1930s these agencies oversaw vast programs of planning and construction of infrastructure, including such major structures as the Hoover Dam.

## *How the Economy Works*

One way to explain economics is to begin with supply and demand and how economies seek equilibrium in prices and the production of goods and services. The concept that illustrates this is a diagram showing the flow of money and labor from firms to workers and back. Workers provide labor to firms, which in turn produce products, which are in turn purchased by workers, who return money for the goods. If a CIS is superimposed on this diagram, then it can be seen that this activity takes place within built environments, the Internet, and transportation flow paths. Also, critical links between the economy and infrastructure systems and

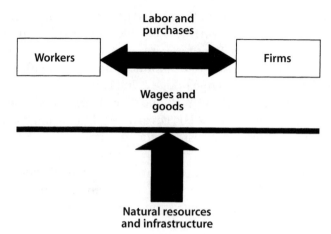

**Figure 10-1** Economic flows, with inputs from natural resources and infrastructure assets.

inputs can be seen (Fig. 10-1), and the flow of needed natural resources to make the economy run can be envisioned. For all this to work, large quantities of raw materials, energy, and water are required (Hyman 1989).

The US economy's mixture of market capitalism and government intervention increases complexity and brings in issues such as regulation, government programs, taxation, and many others. The standard economics diagram that shows markets for wages and goods does not illustrate very well the sectors in which most civil engineers work.

Economics can be divided into microeconomics and macroeconomics. These deal with such small-scale issues as individual choices and with such aggregated sector issues as national level demand and production and related factors (e.g., inflation, unemployment). Microeconomics would seek to explain, for example, how homeowners make choices about using public services, whereas macroeconomics would explain the aggregate statistics of the homebuilding and construction industries.

## Government in the Economy as Regulator

Much of civil engineering work is created by the need to respond to regulations. The government regulates many economic and social spheres, including health, the stock markets, interstate and international trade, the environment, and even educational attainment. The government's reach extends far and wide; how far it reaches is, of course, a controversial political topic. Insofar as civil engineers are concerned, government regulations dealing with infrastructure sectors, with the construction industry, and with employment conditions are perhaps of most interest. For example, each infrastructure sector is regulated or guided by one or many federal statutes (Table 10-1).

**Table 10-1** Examples of infrastructure sectors and statutes.

| Sector | Statute |
| --- | --- |
| Transportation | Transportation Equity Act |
| Water | Safe Drinking Water Act |
| Energy | Federal Power Act |
| Built environment | Occupational Health and Safety Act |
| Waste | Resource Conservation and Recovery Act |
| Communication | Federal Communications Acts |

**Table 10-2** Examples of infrastructure sectors and government services.

| Sector | Government Service |
| --- | --- |
| Transportation | Public roads |
| Water | City water supply utility |
| Energy | Public power utility (e.g., Tennessee Valley Authority) |
| Built environment | Public buildings |
| Waste | City solid waste collection |
| Communications | Startup of Internet by National Science Foundation |

The extent of the government's regulatory activities is an important issue in economics. There are frequently calls for regulatory reform, which often refers to the desire to limit government regulation. The level of regulation is an expression of the balance point in the public's desire for security versus freedom.

## Government in the Economy as Service Provider

The government as a provider of services is a different topic than regulation and deals with the government actually providing such public services as transportation, water supply, electric energy, buildings, or waste collection (Table 10-2).

Government provision of public services is a rapidly changing arena because of trends toward privatization. Actually, three methods of delivering public services are in use—the traditional public model, privatization, and managed competition. In a paper by Geoff Greenough, Commissioner of Works in Moncton, New Brunswick, and colleagues, for example, St. Paul, Minnesota, was presented as the "traditional model" of public services delivery, meaning that public services are provided by city government agencies (Greenough et al 1999). St. Paul is adapting beyond traditional approaches to better meet public expectations. The only thing that seems constant in St. Paul is change, and the situation seems summed up by the phrase "after 10 years of doing more with less, the city is continuing to do more with less" (Greenough et al 1999).

In the paper, privatization referred to turning over services completely or partially to the private sector (Greenough et al 1999). Managed competition might be viewed as a compromise approach that maximizes advantages and minimizes disadvantages. It allows government services, privatization, or a mixture of the two (e.g., when government departments bid against private companies). Charlotte, North Carolina, for example, has found that it improved the city's approach to public service delivery and now allows the city to balance limited revenue sources against costs and demands for services. All three approaches to delivery of public works services offer benefits and challenges. Management trends offer ways to improve by introducing total quality approaches; and in any approach to delivering public works services, the goals are to improve service and lower cost.

## *How Economic Trends are Measured*

As civil engineers encounter business-related economic issues, such as economic efficiency and growth in the economy, indicators such as the GNP will be noticed. A few indicators and measuring tools are essential for understanding business cycles, recessions and expansion periods, rise and fall of the stock market, and other events. A few of the critical indicators are the GNP and gross domestic product (GDP), inflation, interest and discount rate, Federal Reserve actions, stock markets and indices, bond market, deficit, and debt.

## *Saving, Investing, and Consumption*

A big issue for civil engineers working on infrastructure projects is to locate the capital to build them. The US construction industry is taking in approximately $750 billion per year in revenues, much of which is new, invested capital. The definition of *capital* is human effort in storage in the form of assets. Funds that are used to build projects are initially savings and subsequently investments. If the investments are in the right projects, the investor builds "muscle." Projects that are useless investments build nothing. Consumption is the reverse of saving. When people buy consumer goods (e.g., theater tickets, food, clothing), they are consuming and not putting money into savings. In economics there is much debate over the merits and demerits of high versus low savings rates. Americans have a lower savings rate compared with some other countries (e.g., Japan).

## *Public Sector Economics*

Civil engineers generally focus on public sector economics rather than on consumer choices such as the purchase of automobiles or televisions. The main topics include the role of government as regulator and provider of services. Chapter 4 discussed in more detail how most civil engineers work either for government agencies at state and local levels or in firms that provide services for the government.

Public sector economics refers to the activities of local, state, and federal governments, which are a large part of the economy. In 1997, for example, the government's share of the GDP was $1,027 billion, of a total GDP of $8,110 billion. State and local government were about two thirds of the total for government. Of course, the budgets of government exceeded these figures, but the GDP figures show direct contributions toward the economic product of government. GDP measures the value added by economic activity. To measure it requires careful procedures by official government agencies, in this case the Bureau of Economic Analysis of the US Department of Commerce. Government's role in the economy includes government agency actions and expenditures, environmental decisions, the agricultural sector, reactions to special interest groups, and the general economics of the government sector. Regardless of the swing toward free market capitalism or socialism, the government will continue to be a large factor.

## *Money and Banking*

The monetary system is a critical factor in the health and prosperity of the nation. Of particular importance is its role in providing currency to enable transactions to occur and its role in stabilization of the economy and in fiscal policy. The Federal Reserve system and the remainder of the banking and monetary system affect the ability to raise money for projects and to finance enterprises. The Federal Reserve system has the power to stimulate or cool off the economy by setting the interest rates at which it lends money to its member banks.

## *Inflation*

Inflation, the change in value of money compared with a fixed standard, is a critical issue in economic decision making. Civil engineers must consider inflation, for example, in making cost estimates of projects. For example, original estimates might be in 1990 dollars, but inflation (as measured by cost indices) may raise the cost of the project by a significant percentage. Cost indices provide a number of different measures of inflation. The government's Consumer Price Index is the most widely quoted measure, but numerous others exist (e.g., *ENR*'s construction cost indices).

## *Employment*

Employment and inflation, which are closely related topics, affect business conditions for the whole economy, including the construction industry. The rate of unemployment is a widely watched barometer of how the economy uses the human resources available to it. Surprisingly, the goal is not 0% unemployment because it is believed that such a rate would overheat the economy. The base level of unemployment as this is written hovers around 4%, which is the lowest level the US economy has reached in the past few decades. In some European countries, unemployment levels may reach around 10%, reflecting their different approach to employment security.

## *Stimulating the Economy and Labor Economics*

It is good to also consider how economics affects our main client, society, which is concerned with such human issues as jobs and income levels. One important topic of macroeconomics is the labor market or the provision of workers to satisfy demands of the public and private economy. One of the subjects of most concern to civil engineers is the construction labor force. In addition to being important in its own right, the construction work force has been used in the past as an economic tool to stimulate the economy.

During slack times, a popular topic of discussion is the government's role in stimulating the economy through Keynesian tools such as government spending, deficits, and investment through government agencies. The result is deficit spending and increases in the national debt. Civil engineers are closely connected with Keynesian economics because of the popularity of investing in infrastructure to stimulate the economy.

In addition to jobs themselves, labor economics is concerned with such HR issues as the workplace, employment, earnings, job classifications, and related topics. Civil engineers deal with these in the course of managing organizations and firms. One example is government regulation of prevailing wage levels through the Davis-Bacon Act, which affects prices and construction estimates for projects. Another issue of concern to civil engineers is the inequity of income (poverty). Although this is particularly problematic in developing countries, the United States is also affected by poverty, which increases concerns about issues such as social justice and environmental equity.

## *International Economic Issues*

Whereas infrastructure problems in the United States, Europe, and Japan are difficult, they become life-or-death matters in developing countries, which have exploding urban populations, debt crises, and other related problems. International trade is a topic of increasing interest and concern to civil engineers, particularly as the construction market becomes global as consulting engineers seek work and cooperate with other countries. In the international sectors, issues of monetary stabilization, promotion of exports, protectionism, and industrial policy are important and affect civil engineers. Other topics civil engineers may encounter include international trade, wealth and poverty, water and sanitation in developing countries, traffic congestion in cities of developing countries, international development banks, and the role of the International Monetary Fund.

## Economics of Infrastructure and the Environment

The two main economic concerns of civil engineers (providing infrastructure and managing the environment) illustrate how economics focuses on resource allocation. In the same way, economics explains issues that focus on allocating society's resources to other issue areas, such as health care, education, and areas of private consumption.

Expanding on Figure 10-1, Figure 10-2 illustrates the connections of different infrastructure elements to the economy, society, and the environment. Figure 10-3 expands on this further, illustrating the many issues relating to infrastructure.

Figure 10-4 is presented to illustrate the many environmental issues civil engineers face. Shown is a watershed that includes many economic and infrastructure features. The drawing is a diagram from a 1950 federal report, showing that infrastructure and environmental issues do not change very rapidly.

The concept of resource allocation draws a great deal of society's attention. Allocating public resources such as tax revenues is one thing; allocating a person's private resources or property (e.g., land or water) is another. Civil engineers get involved in both types of allocations, requir-

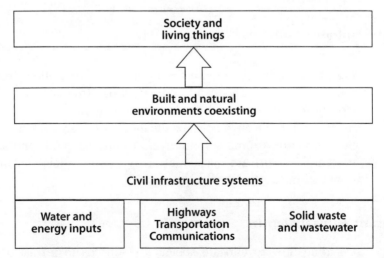

**Figure 10-2** Civil infrastructure systems in support of society and the environment.

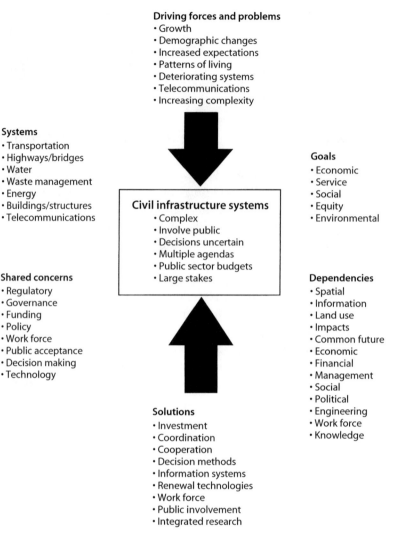

**Figure 10-3** Infrastructure issues at a glance.

ing many interactions with the public. Resource allocation has evolved into public and private decision making. In many ways these relate closely to socialism and capitalism, but there are also many shades of meaning in the terms. Perhaps the basic issue is whether resource allocation decisions are made by individuals buying goods and services in the marketplace (private decisions) or by public officials exercising institutional authority (public decisions).

## Economics of Infrastructure

Deciding how to allocate society's resources to infrastructure systems involves at least two levels of choices. First, how much of society's

**Figure 10-4** A watershed. (Courtesy Corps of Engineers)

resources should go to public goods (e.g., roads, health care, defense, welfare) and how much should be left for private consumption of automobiles, entertainment, clothes, and other private matters? This question is decided in the United States and other free nations by government decisions about tax rates and budgets. Next, how much of the public component of resource allocation should go to capital investment, operation, and maintenance of infrastructure and how much should go to other public programs such as government pensions and aid to education? These are complex questions, but the issue is even more complex than these questions imply. For example, what parts of infrastructure constitute public goods versus private goods? In a transportation system, some choices are about public issues (e.g., resurfacing of roads). The result is a complex mixture of economic issues relating to infrastructure.

Infrastructure issues are closely related to the interdependent economic–social–physical nature of urban systems. Economics seeks to explain complex phenomena such as economic growth, location of jobs and people; economic analysis for urban problems, poverty, housing, transportation, pollution and the urban environment, crime, and urban public services (Schreiber et al 1971).

## *Infrastructure Definition*

Those who write about infrastructure usually devote several paragraphs to explain what it is, and policy reports vary in the categories that are included (Stone 1974; Grigg 1988; Hudson et al 1997). The dictionary definition of *infrastructure* is "the basic facilities, equipment, and installations needed for the functioning of a system" (*Webster's II* 1984). Engineering definitions normally add a conceptual level to this by specifying what the systems do and indicating that infrastructure is physical assets arrayed in systems that provide essential public services. Such definitions include three conceptual ideas about infrastructure: what it is, how it is organized, and what it does. The usual systems included in the definition are those that provide transportation, communications, water, energy, waste management, and the built environment. These perform different functions, and infrastructure is thus a very general concept.

Another issue is that definitions sometimes specify that infrastructure involves only public facilities. This blurs an already confused public–private dividing line and creates a dilemma in discussing privatization. There are two basic ways to clear up the confusion. One is to adopt, as some have, the modifier "public works." The other is to use a definition that relates the built environment to its flows (transportation and communication), its basic inputs (water and energy), and facilities to handle its waste products. Built environments (e.g., cities separated by rural areas or city–suburb complexes) are connected by networks that provide transportation and communications. They are also supplied by water and energy systems, and they generate waste streams that must be processed. These systems characterize infrastructure.

## *Infrastructure Systems Matrix*

With this definition the resulting array of systems and components can be shown by the infrastructure systems matrix (Table 10-3).

The infrastructure systems matrix shows five levels of infrastructure systems, each more disaggregated than the previous one. The most general level is the CIS, consisting of the six systems shown in level 2. CIS is

**Table 10-3** Infrastructure systems matrix.

| | | Systems Level | | |
|---|---|---|---|---|
| 1 | 2 | 3 | 4 | 5 |
| Highest-level system | Industry | Subindustry | Engineered system | Components |
| Civil Infrastructure | Transportation | Highways and roads | Highways and bridges | Bridges |
| | | Rail transportation | Intercity rail | Rails |
| | | Air transportation | Airlines | Hangars |
| | | Water transportation | Waterways | Locks |
| | Communication | Telecommunications | Telephone systems | Cable systems |
| | | TV and radio | Cable television | Cable systems |
| | | Internet | Backbone networks | Fiber cables |
| | Water | Water supply | Water sources | Wells |
| | | Wastewater | Collection systems | Pipes |
| | | Storm water and flood | Collection systems | Inlets |
| | Energy | Electricity | Generation | Turbines |
| | | Gas | Gas production | Wells |
| | Waste management | Solid/hazardous waste | Waste collection | Equipment |
| | Built environment | Residential | Housing | Foundations |
| | | Private nonresidential | Shopping malls | Buildings |
| | | Public structures/facilities | Schools | Buildings |

the conceptual framework developed by the NSF. The NSF explained that CISs are "the constructed physical facilities which support the day-to-day activities of our whole society, and provide the means for distribution of resources and services, for transportation of people and goods, and for communication of information" (NSF 1995).

Each of the six systems at level 2 comprises a large and complex industry, and infrastructure can be thought of as an umbrella or a coalition organizing concept for them. Each of the six subsystems of CIS has at least three subsystems at level 3 that are complex enough to have their own government agencies, regulatory laws, and trade associations. Level 3 can be disaggregated into subsystems at level 4 that are still complex enough to be considered systems and to be divisions of agencies and associations and the subjects of textbooks and annual specialty conferences. Level 5 breaks these subsystems into elements that are, for the most part, components rather than systems. For example, the water system can be disaggregated to water supply, treatment, and distribution; then filters are components of treatment. Table 10-3 reveals that although engineers are normally working on individual components, it is equally important to keep the big picture in mind.

## *Economic Issues*

The infrastructure management arena faces many economic issues (see Fig. 10-3). Driving forces include growth, demographic changes, increased expectations of people, patterns of living, deteriorating systems, influence of telecommunications, and increasing complexity of systems. How does society decide to invest resources in transportation, water, energy, communications, built environment, and waste management? This question has been answered in the past two decades, when much attention has been focused on society's investment in infrastructure. It became clear that increasing demands and past neglect created enormous capital investment needs. This news is not all bad for civil engineers, of course, because fixing the problems increases business opportunities and employment.

In 1981 two economists Pat Choate and Susan Walter used federal reports on facility needs to write a booklet about underinvestment in infrastructure titled *America in Ruins: Beyond the Public Works Pork Barrel* (Choate and Walter 1981). The media picked up the theme of this booklet, and by 1982 there were several national cover stories about infrastructure. Because of these initial spectacular media stories, needs estimates have settled down some but they still amount to more than one trillion dollars for a 20-year period, depending on how they are estimated.

An infrastructure bill (of one trillion dollars over 20 years) would not look so large. For the United States, it would amount to less than $200 per year per capita—not much compared with the total wealth of the country. The total capital stock of the United States varies, but is in the range of $20–30 trillion. For a population of 270 million, that comes to about $100,000 per capita. Much of the wealth is in structures, equipment, housing, and land, and certainly more than $200 per year per capita could be devoted to upgrading and replacing infrastructure.

Many CIS issues need attention. As Figure 10-3 shows, these issues are complex and each involves public decisions. In addition, outcomes are uncertain, multiple agendas arise, public sector budgets must be approved, and the stakes are large. An example of an infrastructure issue, the replacement of an old bridge, is shown on Figure 10-5.

Infrastructure was previously defined as the physical systems that provide transportation, water, buildings, and other public facilities needed to meet social and economic needs. These facilities are needed by people regardless of their income levels. When infrastructure is not present or does not work properly, it is impossible to provide such basic services as food distribution, shelter, medical care, and safe drinking

**Figure 10-5** An old bridge in need of attention.

water. Maintaining infrastructure is a constant and expensive process that is often neglected in favor of more attractive political goals. Economics plays a part in deciding how to allocate society's resources between these competing choices.

## Economics of the Environment

Applying the basic definition of economics to the environment enables us to see the fundamental question: How should scarce environmental resources be allocated? This question would seem arrogant to some because the thought of using something mundane (such as the study of economics) to allocate something so precious as the environment seems wrong. If the human race is to survive, it needs water, shelter, energy and the resources that the environment provides. The central question of environmental economics is how to preserve the environment and at the same time use its resources to support the human race and wildlife?

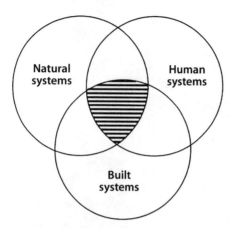

**Figure 10-6** Interaction between natural, human, and built systems.

The link between environment and infrastructure is shown in Figure 10-2 and illustrates how civil systems support both the built and natural environments, which coexist and provide the sustenance needed for society (people) and living things (all other plants and animals). An alternative view of this is shown by Figure 10-6, using a concept produced by the NSF to illustrate the need to work out issues between natural systems (mainly environment), human systems (society), and built systems (mainly infrastructure). Natural systems consist of air, land, and water; and environmental economics deals with how they interact with the other two systems.

## *Sustainable Development*

The concept of sustainable development has been developed to explain the balance in use of environmental resources. Basically, the term means to use resources today in ways to preserve them for tomorrow. Sustainable development is the organizing concept for environmental economics today. The World Commission on Environment and Development defined sustainable development as a process that "meets the needs of the present without compromising our ability to meet those of the future" (Environmental and Energy Study Institute Task Force 1991). It recommended policies on population and human resources, food security, species and ecosystems, energy, industry, and the urban challenge.

In 1994 the President's Council on Sustainable Development had "committed itself to move the United States toward sustainability by the year 2050." It defined sustainable development as "to meet the needs of the present without compromising the ability of future generations to meet their own needs." This was explained further by a group organizing a seminar on sustainable development in the Colorado River Basin: "a sustainable United States will have an economy that equitably provides opportunities for satisfying lifestyles in a safe, healthy, high-quality environment for current and future generations" (President's Council on Sustainable Development 1994).

The term *sustainable development* has attracted a number of definitions and encompasses many different concepts. Some of these definitions and concepts as published by the *Woodlands Forum* include the following (Center for Global Studies 1993):

- Sustainable development is a form of smart growth that employs the high-tech revolution and economic restructuring to manage all this growth in a more sophisticated manner that is ecologically benign.
- Produce nothing until we have a way of fully integrating the product, its by-product, and any by-product from the production process into the system in a positive manner.
- The basic principle of sustainable development is that we live from flows not from stocks.
- Sustainable development is the proper integration of environmental concerns into the development process.
- In a perfect world, an end service such as social security would be sustainable (i.e., there is a circular pattern of continual renewal).
- Pass on to each generation a population level, a set of technologies, and a stock of fertile land and fossil fuels that would enable them to do at least what we have done.

In 1998, the work of the President's Council on Sustainable Development led to a further project to develop a set of sustainable development indicators (SDIs). This is a very interesting development in the sense that the indicators represent the set of measures considered by an interagency group to be the most representative ways to measure sustainable development. The group recognized that indicators should be presented in economic, environmental, and social categories. After reviewing the definitions and concepts for sustainable development, they decided to adopt a pressure–state–response framework to explain indicators and an additional framework to organize the indicators according to whether they are long-term endowments and liabilities, processes, or current results. Pressures are human activities that affect the environment, states are states of the environment or natural resources, and responses are societal responses to environmental concerns. This model, however, does not adequately recognize economic and social issues. For civil engineers, long-term endowments and liabilities include such categories as public infrastructure, natural resources stocks, and environmental conditions. Processes include generation of pollutants and decision making about investments. Current results include items such as low pollution, recreational opportunities, and safety. The group selected 40 indicators, organized according to 20 issue areas in three categories (Table 10-4; US Interagency Working Group on SDIs 1998). Because civil engineers are so closely identified with sustainable development, they should be able to explain each of these indicators and how infrastructure and environmental projects relate to them.

## *Environmental Decision Tools*

For some areas of economic analysis, quantitative tools such as benefit–cost analysis can be used. In environmental economics the issues include so many intangibles that other techniques (e.g., environmental impact assessment) are needed. In environmental impact assessment, including the preparation of environmental impact statements, the task is to inventory environmental resources such as plants, animals, water quality, air quality, and visual amenities and assess how a proposed action will affect them positively and negatively. There are, of course, many other things to say about environmental economics. The topic merges immediately with environmental politics because of the tension between private market and government solutions to problems of finding the way in a regulatory thicket.

# Engineering Economics

Civil engineers should acquire skill in the set of tools known as engineering economics. This involves the use of such quantitative techniques as cash-flow analysis, benefit–cost analysis, and discounting of money to discern the advantages and disadvantages of various courses of action involving capital investments, whether in the private or public sector.

Although there are different applications in engineering economics, a few key ideas can go a long way. Starting with simple notions of compound interest, you can analyze many different situations. For example, the basic equation of compound interest, $F = P*(1+i)^n$, enables the computation of the future sum of money, compounded for $n$ periods at interest rate $i$, based on a present sum of money, $P$. For example, the accumulated money in a savings account that began with $1,000 that was left in the account for 25 years at 5% interest per year is $3,386.35.

Somewhat more complex is the formula for a series of payments, in which there are equivalent sums of money spread over different time intervals. These formulas are programmed on spreadsheets, and they enable quick computation of many useful quantities. Civil engineers need a good understanding of engineering economics, and a number of practical textbooks are available on the subject. Computations in courses in engineering economics and finance are quite similar.

**Table 10-4  Sustainable development indicators.**

| Issue | Selected Indicators | Category |
|---|---|---|
| Economic prosperity | Capital assets | Economic |
| | Labor productivity | Economic |
| | Domestic product | Economic |
| Fiscal responsibility | Inflation | Economic |
| | Federal debt: gross domestic product ratio | Economic |
| Scientific/technologic advancement | Investment in research and development as a percentage of gross domestic product | Economic |
| Employment | Unemployment | Economic |
| Equity | Income distribution | Economic |
| | People in census tracts with 40% or greater poverty | Social |
| Housing | Homeownership rates | Economic |
| | Percentage of households in problem housing | Economic |
| Consumption | Energy consumption per capita and per $ of gross domestic product | Economic |
| | Materials consumption per capita and per $ of gross domestic product | Economic |
| | Consumption expenditures per capita | Economic |
| Status of natural resources | Conversion of cropland to other uses | Environmental |
| | Soil erosion rates | Environmental |
| | Ratio of renewable water supply to withdrawals | Environmental |
| | Fisheries utilization | Environmental |
| | Timber growth to removals balance | Environmental |
| Air and water quality | Surface water quality | Environmental |
| | Metropolitan air quality nonattainment | Environmental |
| Contamination and hazardous materials | Contaminants in biota | Environmental |
| | Identification and management of Superfund sites | Environmental |
| | Quantity of spent nuclear fuel | Environmental |
| Ecosystem integrity | Acres of major terrestrial ecosystems | Environmental |
| | Invasive alien species | Environmental |
| Global climate change | Greenhouse gas emissions | Environmental |
| | Greenhouse climate response index | Environmental |
| Stratospheric ozone depletion | Status of stratospheric ozone | Environmental |
| Population | US population | Social |
| Family structure | Children living in families with only one parent present | Social |
| | Births to single mothers | Social |
| Arts and recreation | Outdoor recreational activities | Environmental |
| | Participation in the arts and recreation | Social |
| Community involvement | Contributing time and money to charities | Social |
| Education | Teacher training level and application of qualifications | Social |
| | Educational attainment by level | Social |
| | Educational achievement rates | Social |
| Public safety | Crime rates | Social |
| Human health | Life expectancy at birth | Social |

## Finance for Civil Engineers

Civil engineers are involved in management more than are other types of engineers. Civil engineers in both the private and public sectors find that finance becomes more important to them as their careers progress. They are not required to master all details of accounting, but they must be able to obtain the revenues they need and manage them. Without adequate funds to build and maintain facilities or operate organizations, success is not possible. This applies to all levels and divisions of organizations, although each job brings a different perspective as a result of differences in budgets and levels of responsibility. It is important that the people who really know what is going on in an organization also understand how the finances are worked out.

## Personal, Business, and Public Financial Management

Financial management has private business, public, and personal aspects. Public finance is a special field dealing with managing money in the public sector. Many engineers work in private finance as they become involved in their own companies or serve private sector clients (e.g., land developers). Many other civil engineers work in the public sector and in the world of public finance. Personal finance is similar to the other areas but involves your own money and personal organization. Lessons learned about personal finance can leverage knowledge about business and public finance.

Personal finance deals with issues such as personal income and expenditures, budgeting, investing, and managing finances. Many issues related to personal finance will be encountered during a career and lifetime. These include making enough money, buying real estate and insurance, obtaining and managing credit, buying capital items such as automobiles, investing, establishing a retirement account, and eventually setting up an estate. Although these are not the subject of this book, each item of personal finance has a parallel in the business world (Table 10-5).

Although the field of business or corporate finance is more complex than is personal finance and includes issues related to financial markets, institutions, securities, risk management, and other topics, it is useful to study the parallels among the three financial sectors. In many ways, public finance is similar to business finance, but it also has links to government functions such as political budgeting, raising public funds, accountability to taxpayers, and public sector accountability. Regardless of the arena of financial management, the central issues are organizing the financial management function, financial planning and budgeting, raising revenues and other funds, planning and managing expenditures, accounting, and reporting.

**Table 10-5** Similarities between personal, business, and public finances.

| Personal Finance | Business Finance | Public Finance |
|---|---|---|
| Income | Sales and revenues | Budget allocation and revenue |
| Budgeting | Budgeting | Budgeting |
| Keeping records | Accounting | Public fund accounting |
| Real estate | Property and real estate | Property |
| Personal insurance | Liability and other insurance | Liability and other insurance |
| Managing credit | Managing credit, issuing stock and bonds | Government fund credits |
| Paying taxes | Managing taxes | Managing tax issues |
| Buying capital items | Capital items | Capital management |
| Retirement account | Employee benefit and retirement programs | Public retirement programs |
| Charity | Corporate philanthropy | Managing philanthropy |
| Estate planning | Mergers and acquisitions; succession planning | Mergers of government entities |

## Organizing the Financial Management Function

Organization of financial functions varies from one business to another. In a business the person who handles finance might be called a chief financial officer, a vice president of finance, a controller (comptroller), or simply a business officer. In a public agency, there might be an assistant city manager for finance and administration, and in a university the finance office might be focused on business and financial services, including tuition, student aid, and other financial issues.

Regardless of the titles, the functions of financial offices include budgeting, accounting, auditing, assessments, purchasing, and treasury. The budget office handles the compilation of budgets from various offices and presents the budget to the chief executive officer, who will in turn present it to the governing board. The accounting office will be responsible for keeping the books and handling the reports. Auditors can be internal or external, responsible for making independent assessments of financial condition. Assessment is a public finance function, necessary to support tax collection. Purchasing controls the physical assets of an organization to make sure it gets good value and that property is tracked and managed. Treasury generally makes collections and disbursements.

## Financial Planning and Budgeting

Financial planning and budgeting indicate the set of activities needed to make an overall assessment of an organization's financial prospects and actions and to make plans to collect revenues, control expenses, and ensure that financial resources are used wisely. Financial planning includes revenue analysis, cost analysis, institutional analysis, ability-to-pay analysis, secondary impacts analysis, and sensitivity analysis (Government Finance Research Center 1981). These analyses provide a framework to consider financial issues, whether they relate to personal, business, or public activity (Table 10-6).

Budgeting, a powerful management tool for personal, business, or public finance, links the planning, operating, and controlling functions of management. A budget is much more than just a tool for allocating money. Planning for the budget, creating and managing work programs based on the projected budget, and evaluating what was achieved from the budget are management tasks that revolve around the budget. Actually, budgeting as we know it today has not been around all that long. The concept is credited to the New York Bureau of Municipal Research, which worked from 1907 to 1915 to reform municipal administration (Grigg 1988).

The budget process is a cycle of planning, negotiating, and implementing a budget. For example, consider the simple case of household budgeting. A family will examine its needs, income, and goals and make a plan for expenditures for items such as rent/mortgage, food, transportation, clothes, and other expenses. A business would do the same thing, but income would be replaced by projected sales and revenues and planning would include other items of fund-raising such as sale of stock, loans, and perhaps mergers and acquisition activity. In public finance, a budget would focus on revenues from taxes, fees, and bond proceeds. Budgeting involves many decisions about the policies and directions of the organization, whether it is a family, business, or public agency. There are two types of budgets: operating budgets and capital budgets. The operating budget refers to items that are short term in nature and is used to keep the unit in operation. Examples are regular income, daily

**Table 10-6** Elements of financial planning.

| Planning Element | Personal Finance | Business Finance | Public Finance |
|---|---|---|---|
| Revenue analysis | Income analysis | Sales | Tax and fee revenues |
| Cost analysis | Personal expenses | Business costs | Government expenditures and budgets |
| Institutional analysis | People involved, reliability, and maturity | Business track record | Public institutions |
| Ability-to-pay analysis | Payments record, overall assessment | Track record, business assessment | Cash-flow analysis |
| Sensitivity analysis | Reliability of forecasts | Reliability of forecasts | Reliability of forecasts |
| Secondary impacts analysis | Other people involved | Effects on other businesses and markets | External impacts of activities |

expenses, utility fees, repairs, and maintenance. The capital budget deals with such long-term assets as real estate, vehicles, and other equipment.

One decision is the level of taxation and charges to the community or the portion of total community resources needed for government programs and services. Another is the emphasis that will be placed in different programs within the governmental structure. Then there are decisions within specific programs of how money will be allocated to personnel, equipment, contracts, and other expenditures. The budget also states how the revenue will be made available, whether from debt, user charges, or other sources. The capital budget should be linked to a comprehensive infrastructure planning and needs assessment process. The operating budget should be linked with plans for services, organizational development, and the development of programs.

The federal budget process is far more complex and political than any other, simply due to the scope of the enterprise. Many interest groups and types of programs compete for funds. Decisions about the federal budget have far-reaching impacts on economic health and even on international matters such as the strength of currencies. The year of budget planning is intense, and agencies battle to submit budget requests. The process culminates with review by the Office of Management and Budget (OMB), which may refuse to allow budget requests to go to the president. This culminates with the president's budget message to Congress in early February. This is only the beginning of the struggle, which lasts until the beginning of the federal fiscal year on October 1 (often longer), with threats to shut the government down and similar political moves.

For any agency, the positive resolution to approve an operating budget (e.g., by a city council) provides an authorization to spend for the organization. The negative side of the budget is its use to contain spending. Uses of the annual operating budget are to provide information to aid in planning; require executives to produce an estimate of expenditures so that the adequacy of revenues can be checked; provide a means to evaluate internal competition for resources; use in work planning and evaluation; communicate with the policy oversight body the operating objectives for the coming year and to make revisions based on signals from that group; provide information for the annual appropriation ordinance; provide a basis by which annual plans can be adjusted to conform to appropriations; and provide a basis for financial audit of the enterprise (Moak and Hillhouse 1975).

Budget politics are of concern to managers at all levels of government, especially those involved in infrastructure where much money is involved. We may look at the politics of budgeting as internal power plays and the aspirations of external constituencies. The internal politics have to do with bureaucrats gaining power, and the external politics have to do with the interest groups gaining programs or influence they seek.

External groups have lobbyists working for them, so the level of conflict rises quickly in this aspect of budget politics.

## Revenue Management

Revenue management means to ensure that the unit has enough money to operate and achieve its goals, whether they relate to personal, business, or public agency situations. The operating and capital budgets should be financed from logical sources. Operating funds should come from current revenues, with minimum reliance on subsidies when they can be avoided and with a close connection between the service rendered and the charge imposed.

In personal situations, revenue management normally means having a good job. Other sources of personal funds include gifts, investment income, and special situations. In businesses, revenues come from sales. In public organizations, they come from taxes and fees. Having a good job relates to the civil engineer's career (see Chapter 3).

In business, having adequate and growing revenue is central. If the business is a consulting engineering concern, for example, the main goal for success is having a client base that produces enough reliable revenues and sustaining and growing the operation. Any business faces this same challenge. It is the goal of business owners and managers to develop product and service lines that meet public needs and lead to a good and growing revenue stream.

In public enterprises, many of the same goals apply: providing a service that produces value for the customer, collecting revenues, and holding costs down. In addition, public finance can be used to illustrate several important points about revenues. Revenue sources that are normally used in public sector operations involving civil engineers are for the development and operation of infrastructure systems. These center on rates and user charges, property taxes, and other taxes. Operating budgets must be renewed each year and should be financed from recurring revenues. In past years the property tax was used to finance operating budgets in many organizations, although with today's emphasis on enterprise budgets, the trend is toward reliance on user charges.

Rates and user charges follow the "user pays" principle, which is easy to accept in charges for such services as garbage collection and electricity. However, it can also apply to situations such as storm water management and streets maintenance. User fees should provide both economic efficiency and economic "equity" in the provision of public services. Efficiency means there is no waste and that the public gets what it pays for and rations use of the public service accordingly. Not wasting electricity is an example. Most people are careful about the use of electricity because that charge is often a significant part of personal budgets; however, people are apt to waste water, even when it is metered, because the unit cost is lower. The theory for setting user fees for electric, gas, and telecommunications utilities is in the field of utility economics. In the case of water, user fees are studied by the American Water Works Association. Roads are studied in other places; and with privatization, there is much interest in financing roadways using means other than taxes.

Opposition to user charge increases is common. An impact fee is a unique charge related to new development and attracts opposition from developers. Governments are not regulated by public utility commissions, and opposition must be registered through political channels. Despite the popularity of user charges, tax revenues remain in use as well, particularly for services for which it is difficult to identify the beneficiaries. Normally, the major source of tax revenues is the property tax, which is called an *ad valorem* tax because it is calculated according to the value of the property. States have different formulas to apply a mill levy against the assessed valuation of property. Other taxes that might find their way into infrastructure revenue streams are income and sales taxes.

Capital financing for infrastructure facilities comes either from current revenues or from debt financing, primarily general obligation or revenue bonds, or from combinations of these sources. Attention is also being given to such nontraditional sources as public–private cooperative ventures. When financing is from current revenues, provision must be made to divert the revenues into a capital reserve account; this is known as pay-as-you-go financing and is similar to saving for a personal purchase. Using debt financing may be more effective, however, because of the equity issues involved in financing long-term facilities. This is similar to borrowing money to pay for an item. Using current revenues means that current-rate payers are paying for the facilities that will be used in the future by others. Debt financing may be more cost-effective because of inflation, uncertainty, and the opportunity to invest revenues elsewhere. If interest rates on debt are low, then it pays to use it to build. If they are high, then current revenues may be a better choice, with the option of borrowing at more favorable terms later.

Debt financing is known as pay-as-you-use financing. If the term of the repayment is the same as the life of the facility, then debt financing provides an exact match between those paying for and those using facilities. Nothing is perfect in this world, of course, and we do not generally know the lifetimes of facilities nor do we know what repairs they will need. Bonds are a common vehicle to finance infrastructure projects. General obligation bonds, backed by the credit of the organization issuing the debt, and revenue bonds, depending on dedicated revenues of a self-supporting project, are the common types. System development charges have become an important part of capital financing strategies because they provide a way to isolate the cost to serve a particular segment of a system and to levy charges for it. In effect, they allow new users to "buy into" an existing system by paying for their fair share of it. A variation of debt financing that also involves public–private cooperation is tax increment financing, defined as an approach that uses the increase in taxes that occurs after a development is finished to repay debt.

Grants have been an important part of the overall strategy of paying for infrastructure systems. The wastewater construction grants program of the 1970s and 1980s, for example, financed some $40 billion in treatment facilities over 10 to 15 years. Intergovernmental revenue, as grants are sometimes called, is an important part of the financing of local infrastructure.

## Cost Control

Cost analysis in financial planning and management also involves operating and capital items. In personal finance, cost refers to the personal or family budget. In a business, it refers to the costs involved in operating and building the business. In a public agency, it refers to the necessary expenditures. Civil engineers encounter construction costs, operating and maintenance costs, and other costs such as for regulatory programs and planning. Cost analysis may involve techniques such as value engineering and others that study how to cut waste in the system. Cost analysis is also important when the financing study determines the components of cost that can be assigned to different users.

Costs can be classified as direct and indirect. Direct costs are those directly assignable to the provision of a particular service. Examples are wages, equipment, operation and maintenance expenses, depreciation, and capital expenses. Indirect costs are those that are necessary for the delivery of a service but that cannot be attributed directly to the service itself. Examples would include central services such as computer and support services. Cost control is a matter of making sure that full value is received for every dollar spent. This is a function of management at all levels and requires careful attention to the planning and approval of expenditures as well as postaudits to determine how well the investments in program and equipment have paid off.

## Accounting

Accounting means to monitor income and expenditures and to maintain records and reports. Accounting maintains information and analysis to provide management and outside interests with the facts needed to make decisions. Personal finance involves simple accounting, but for businesses and government organizations accounting is much more complex. At the highest levels, accounting includes performance assessment as well as tracking money. An example can be seen in the activities of the US Government's General Accounting Office (GAO).

The value of accounting can be seen from two types of reports: the income statement and the balance sheet. Income statements provide an estimate of the differences between revenues and expenditures over a period of time (e.g., 1 year). For example, in personal finance or in a small business, the income statement would show whether there were profits or surpluses in a year or whether the year ended in a deficit. In private business this is sometimes called the profit and loss statement and reports how well the enterprise did during the previous fiscal year.

The balance sheet provides a report of changes in assets and liabilities over the accounting period. It is a cumulative snapshot of the financial picture of an accounting unit, whether an individual, a business, or a government agency. The balance sheet shows how the finances balance at an instant in time. The balance sheet reaches a balance by including earned surplus (or net worth) on a statement.

The income statement and balance sheet require the use of accounting values such as depreciation and book value and are more the province of the accountant than the manager. For the manager, however, the income statement will be a financial control device, with the main interest being in cash receipts and cash disbursements, and the balance sheet will mainly be a way to display the debt structure and the cash accounts payable. Financial statements can be compared to statements of water balances in a reservoir. The statement of revenues and expenses is like the annual water budget where the report is of inflows, outflows, and change in storage. The balance sheet is like the report of how much water is in the reservoir at the end of the year along with how much is owed to users and how much is expected from others.

Civil engineering students often lack exposure to and understanding of basic accounting statements, and this is an area in which they should give more study. The manager can also take advantage of the publications and guidelines of the professional groups that support the fields of finance. For example, the Government Finance Officer's Association has good publications and services.

## Financial Control and Reporting

Financial control is accomplished through accounting and auditing, with its reports and checks and balances. Reports furnish the management information needed to make adjustments and to report to boards of directors, customers, and regulatory agencies. *Auditing*, a term generally included in the more general term *accounting*, is the process of examining accounts or making an outside check on the validity of the financial management and the health of the enterprise. As a regulatory measure, auditing is generally carried out by accountants other than those who keep the regular accounts.

The key report of the organization is the annual financial report, which contains the overall results of the previous year's activities, including operational and fiscal performance. An effective annual report can focus management attention on the results achieved and on the financial health of the organization. This is easily seen in private organizations in which one of the primary goals has been the bottom line, with the stockholders holding management accountable for profits and share appreciation at the annual stockholder's meeting.

In a public enterprise the goals are usually more complex and politically set than the profit goals of a private company, and it is not normally as straightforward to report results and financial position. Goals of most infrastructure organizations can be quantified in terms of such indicators as number of vehicles handled, gallons of water delivered, and tons of trash handled, with financial results reported in terms of service delivered as a function of cost. These are performance indicators that might be used in the annual report.

How financial statements are presented is a decision of management, with guidance from sources of accepted practice in the financial management profession. The National Council on Governmental Accounting has issued guidelines, one of which specifies a comprehensive annual financial report, which includes the elements of financial reporting most necessary for management and outside interests. Financial control involves more than accounting. The built-in checks of the organization for control of purchasing, the fixed assets records, inventory, and the control of hiring through staff activities all support management of the financial health of the organization.

The GAO, part of the Office of the Comptroller General of the United States, oversees programs financed with federal money, an enterprise of about two trillion dollars per year. To extend the traditional financial audit into the broader function of performance evaluation, the GAO has begun to use the term *performance audit* to include financial, economic, and programmatic elements. The definition of these three elements by GAO is "financial and compliance and determines (a) whether financial operations are properly conducted, (b) whether the financial reports of an audited entity are presented fairly, and (c) whether the entity has complied with applicable laws and regulations" (US GAO 1982).

## Public versus Private Finance

Many civil engineers are involved with some aspect of public finance or management of the public's money for purposes such as infrastructure systems or environmental protection. In public finance, it is important to understand the difference between self-support, or the enterprise principle, and dependence on subsidies. Public works management should be on a self-supporting basis, following the "enterprise principle." The concept is that services should be self-supporting and charged according to the benefits users receive from the services. The use of pricing through user charges is the basis for control of the allocation of services and for raising revenue. The equity issue is central to the philosophy that the charging schemes should be fair.

If a service is self-supporting, revenue generation and financial control are brought under the control of the manager rather than the political process. This is the major advantage of the enterprise approach, in which managers gain control to make decisions and implement innovations in financial management. Still, there is a built-in political component in infrastructure management and the desires of the local population must be factored into decision making. This is one of the principal issues in whether to privatize a public service. The financial management process must consider both political and administrative factors. This is healthy and represents a good balance between bureaucracy and politics.

The manager of a self-supporting organization will have the entire organization in mind when considering financial aspects. If the organization is large and complex it will be divided into subdivisions, each a center of responsibility and possibly a cost center. If it is a cost center, then there will be separate budgeting, control, and financial responsibility. Examples of cost centers might be a full water utility, a transportation department or division of a local government, or a project that is determined to be a cost center because of the need for it to break even on the enterprise involved.

Subsidies cannot always be avoided, as in the case of providing vitally needed services when citizens cannot pay for them. The use of subsidies for transit is common, for example, because the fares paid by bus or tram riders do not pay the full cost of operating the systems. The federal government has been providing operating subsidies for transit on a general basis for a number of years. Other examples of subsidies are in the construction grants program for wastewater and the construction and operation of public housing. The use of subsidies in developing countries is widespread, often providing the difference between life and death. In the case of irrigation systems in developing countries, for example, even though operation is not directly subsidized it is indirectly subsidized through the allowance of deferred maintenance, with catch-up grants and soft loans for rehabilitation.

# 11 Law for Civil Engineers

## How Law Impacts Civil Engineering

As our society becomes more litigious, people are more likely to use the legal system to resolve disputes than they are to work them out through voluntary cooperation. This raises conflict levels and necessitates that civil engineers be concerned with legal procedures in all business and personal matters.

Legal issues top the list of greatest concerns for civil engineers, especially engineers in business and management. A 1999 survey by the Denver-based Janine Reid Group confirmed that the issues of most concern to civil engineers are construction delays, design errors and issues, cost overruns, management succession, neighborhood opposition to a project, employee raiding by competitors, third-party lawsuits, disgruntled employees, mergers and acquisitions, and accidents involving a company lawsuit (ACEC of Colorado 1999). The fact is, most civil engineering issues involve law to one extent or another (Table 11-1).

In addition to the issues listed in Table 11-1, civil engineers routinely encounter many other categories of law, administrative procedures, and regulatory controls. Civil engineers often find that navigating in legal channels is as important as navigating the regular technical work. Engineers do not need to be legal experts or know how to practice law (lawyers should be called for complex matters); however, they do need to know some aspects of law.

This chapter presents an overview of the law as we have experienced it (i.e., as nonattorneys). This chapter does not provide legal advice (or

**Table 11-1** Impact of law on civil engineering issues.

| Civil Engineering Issue | Type Of Law |
|---|---|
| Construction delay | Construction |
| Design error/issue | Construction/design |
| Cost overrun | Business/financial |
| Management succession | Corporate |
| Neighborhood opposition to a project | Environmental or other law |
| Employee raiding by a competitor | Business |
| Third-party lawsuit | Civil lawsuit |
| Disgruntled employee | Employment/labor law |
| Merger/acquisition | Corporate |
| Accident involving a company lawsuit | Personal injury/civil lawsuit |

even expert views) on matters of law; rather it provides an engineer's perspective on law that is important in the areas in which engineers work. For expert views on law, readers are advised to consult an attorney.

## Scenarios

The range of situations in which civil engineers may encounter legal dilemmas is broad. Law that relates to design and construction is paramount and involves contracts, specifications, and liability issues. The examples presented in Table 11-2 illustrate different types of law related to the roles played by civil engineers. The list is not exhaustive and does not represent all categories of law that may be encountered by civil engineers, but it does present ranges of situations that illustrate what civil engineers are likely to face.

## Framework for Law

Law can be classified in many ways, such as by civil or criminal law, by levels of government, and by types of legal instrument (constitution, statute, regulation, executive order, and case). In addition, international law is becoming increasingly important in the global business environment. An outline of the legal field includes the following topics (Burnham 1995):

- Evolution of law in the United States
- Legal methodologies
- The adversary system and jury trials
- The legal profession
- The judicial system

**Table 11-2    Example civil engineering scenarios and legal situations.**

| Sector | Level and Role | Task | Legal Situation |
|---|---|---|---|
| Water | Consulting engineer | Plan improvements to water treatment plant | Compliance with the Safe Drinking Water Act and rules of the Environmental Protection Agency |
| Transportation | Consulting engineer | Design highway bridge built over a river | Use of rules of highway acts to set parameters for the bridge and consideration of liability in dealing with flood parameters |
| Built environment | Field engineer | Make decision that construction of project must halt until changes are made | Know legal authorities to protect client when contractor threatens to sue if the project is halted |
| Environment | Regulator in state agency | Assess permit application | Know legal processes mandated by environmental laws that must be followed for project to proceed |
| Transportation | Engineer for state department of transportation | Intelligent Transportation Systems | Have thorough knowledge of regulatory framework for highway safety and capacity |
| Wastewater | Public works or utility manager | Head a wastewater improvement project planned by the city | Know Clean Water Act |
| Structures | Specialist in pipe company | Evaluate structural strength of a pipe design | Know codes for pipe strengths |
| Built environment | Primary/subcontractor: engineering firm acts as subcontractor to primary firm | Prepare subcontractor specialty report (e.g., soils report) | Be familiar with legal liability of all parties involved in building foundation |
| Water | Expert witness | Analyze flood event | Know laws such as those that authorize flood control projects by the federal government |
| Energy | Business partner | Internal presentation to management on potential power plant project | Be able to interpret to management the risks and opportunities in all business aspects of the potential project |
| Environment | Public works director | Public involvement for a hazardous waste facility | Know public rights under hazardous waste statutes |
| Water supply | Utility engineer | Emergency response to contamination incident in water system | Know emergency response and police laws such as the responsibility of public agencies to handle emergencies |

- Administrative law
- Civil procedure
- Criminal procedure
- Constitutional law
- Contract law
- Tort law
- Property law
- Family law
- Criminal law
- Business law

In addition, other categories of law that are especially important to civil engineers include professional practice, construction and design, engineering contracts, environmental and water law, and interstate law.

## Evolution of Law in the United States

Law in the United States evolved along with the governmental structure initiated by the Constitution in 1789 and its predecessor institutions. To function in society, it is important to understand the basic structure of government, legal rights of citizenship, and other important issues about civic duty and behavior. The origins of law in the United States can be traced to such early civilizations as the Greeks, Romans, and Hebrews. Moses recorded the Ten Commandments in the Old Testament as one of the earliest statements of a code of behavior. Civilization took a major leap forward when it moved from a system in which disputes were settled only by force to a system in which rules prevail (rule of law). Rule of law evolved in medieval Europe, and early settlers brought with them the English common law, from which US law evolved. Other doctrines of law, such as the Napoleonic Code, also influenced law in the United States.

## Legal Methodologies: The Adversary System and Jury Trials

The law generally functions within a judicial framework, using the adversary system, jury trials, cases, and presentation of evidence by opposing sides. These terms have become familiar because so many instances of police action, courts, trials, and jails are portrayed in movies and on television. Movie and television portrayals often focus on criminal law, although some television programs feature civil law issues, such as personal conflicts between people. The principles of both criminal and civil law are similar; but the burden of proof is different.

The adversary system means that the sides take opposing views. The plaintiff in a civil case is suing the defendant. In a criminal case, the prosecuting attorney attempts to prove a case against an accused defendant. Similar to a debate, the adversary system sharpens both sides. Each side must carefully examine its facts and opinions, compile them, and then present them in a coherent manner to a judge and/or jury or resolve them before trial. Much law operates outside this judicial framework, although the framework may still supervise such legal issues. If every issue had to go to a jury trial, society could not function very well.

## *The Legal Profession*

Lawyers are the main element of the legal profession, and they carry out the main work of the legal system. The legal system also includes police, judges, administrative officials, and other support personnel. Because lawyers are the driving force in the legal system and are so prominent, they attract public attention and comment.

The legal profession has similarities to engineering, such as a professional education, a system of ethics, work on a fee basis, and professional licensing. Engineers often employ lawyers to assist them and vice versa, and engineers work closely with lawyers on joint projects. Similar to engineering, law has many branches, including admiralty and maritime law, civil litigation, entertainment law, international law, labor law, natural resources and environmental law, sports law, family law, public interest law, government law, small-town practice (general practitioners), employment law, and personal injury law. Engineers may encounter several areas of law over the course of a career. For example, one of the authors has been an expert witness in a maritime case, a defendant in civil litigation involving employment law, and involved in environmental law.

Lawyers are also known as attorneys, counselors, or solicitors. In the United Kingdom, a lawyer is called a barrister. Law school usually lasts 3 years, with graduates earning a doctor of jurisprudence (J.D.) degree. States require lawyers to pass a state bar examination to be allowed to practice. This is the equivalent of the professional engineer registration for civil engineers. The United States has 183 law schools accredited by the American Bar Association. In 1998, approximately 680,000 lawyers were in legal practice, and there were approximately 70,000 judges and other judicial workers. Today approximately 750,000 graduate lawyers are in the work force, about half the number of graduate engineers; there are twice as many engineers as lawyers (US Department of Commerce 1998).

Engineers often work with attorneys on projects and problems related to policy, potential or ongoing litigation, and contract issues. Figure 11-1 shows legal work taking place in a law office. It is interesting to note the many volumes of statutes and cases in the bookcase depicted in this figure. Engineers often will need to consult law books such as these. Civil engineers can often benefit by learning to navigate among the sources of information available to the legal profession.

**Figure 11-1** Attorneys in a law office. (Courtesy Environmental Protection Agency)

## *The Judicial System*

Much of the legal process in the United States takes place within the judicial system, which consists of trial and appellate courts at the local, state, and federal levels, culminating with the US Supreme Court. Federal and state courts derive their authorities from their constitutions and statutes. Each system handles different categories of criminal and civil cases. Federal courts try cases involving federal law and cases between parties from different states or from other countries. The federal court system includes district courts, courts of appeals, and the Supreme Court. District courts hear original cases. There are 95 district courts in the United States, with each state having at least one. There are also circuits (districts), each of which has a court of appeals. Additionally, the US Court of Appeals for the Federal Circuit covers the nation. The Supreme Court of the United States, the highest court, may hear cases appealed from a federal or state court of appeals, but it carefully chooses the cases it will hear. The US Supreme Court also has original jurisdiction over cases involving two states or representatives of other countries. There are also several specialized courts, including the US Claims Court, the Court of International Trade, and military courts.

State courts handle criminal and civil cases and include police courts, magistrate's courts, county courts, and justices of the peace. There are many other types of courts, including small-claims court, probate court, courts of domestic relations, juvenile court, and traffic court (*World Book Millennium 2000* 1999).

## *Administrative Law*

Civil engineers encounter administrative law more than other kinds. This is the set of rules developed by administrative agencies, such as the EPA, to carry out their statutory authorities. This authority is undertaken by rulemaking (i.e., the development and issuance of regulations that spell out how statutes are to be implemented). Agency officials have a great deal of power in rulemaking and in judicial decision making. There is judicial review of agency decisions, meaning that any decision made by an administrator is subject to review in a court of law. Administrative law is also found in the operation of such government functions as banking, communications, trade, transportation, education, public health, and taxation as well as any area in which government functions.

The United States has an Administrative Procedures Act, which serves as the organizing code for rules about procedures. Code topics include administrative agencies' rules and practices, administrative courts, boards and commissions, administrative regulations, administrative remedies, complaints, judicial review of administrative acts, licenses, administrative sanctions, and tax courts. Code topics also include organization of and procedure before special courts or administrative (quasi-judicial) agencies concerned with the adjudication of cases such as taxes and revenues.

## *Civil Procedure*

Procedures in the legal system are generally known as civil or criminal. Civil procedure involves cases in which one party (plaintiff) sues another (defendant) in a municipal, state, or federal court. Cases go through stages of pleas before the court, including discovery, motion for summary judgment, pretrial conference, trial, and judgment. Most legal situations faced by engineers will be concerned with civil law. Cases may be settled by negotiation rather than with litigation. One such negotiation process, alternative dispute resolution, is common in cases involving construction disputes.

## *Criminal Procedure*

Criminal procedure is familiar to anyone who watches movies or television because it is so prevalent in the media. Stages include investigation, arrest, formal charges, first appearance in court, indictments, pretrial motions, trial, sentencing, appellate review, and incarceration. It is hoped that civil engineers will not be involved in criminal procedures, but they are occasionally necessary in the course of work. For example, if a business executive is charged with criminal acts in the course of a pollution incident, the executive could be convicted as a criminal and go to jail. Civil engineers might be involved in the investigation or testimony of such a trial.

## Constitutional Law

Constitutional law refers to legal questions, procedures, and actions taken under the Constitution and deals with issues such as judicial review, separation of powers, the federal judiciary, issues between Congress and the president, state–federal relations (federalism), relationships between states (e.g., interstate compacts), due process, equal protection under the law, freedom of speech, and other issues. Engineers are normally not concerned with constitutional law, but it does serve as an organizing framework over other categories of law that do concern engineers.

## Contract Law

Engineers need to use contract law regardless of the category of their work. Common uses of contracts that involve engineers are construction contracts and contracts to do engineering work. Contracts are integral to business and the function of public sector activities, and they specify the agreements of the parties to carry out allocated duties, responsibilities, and obligations. Most engineers involved in consulting engineering work, construction, and public works management are faced with many instances of contract negotiations and administration. Some categories of contract law include purchase contracts, sale contracts, buyout agreements, prenuptial and postmarital agreements, wills, trusts, employment contracts, property settlement agreements, enforcement of contracts, and leases.

Much of the law involved in construction is about contract compliance. For example, in one case a contractor lost an appeal over payment delays. The decision of the US Court of Appeals for the Eighth Circuit was in favor of the owner, reversing a jury award of $21.5 million for the contractor in District Court. The Court of Appeals ruled that the owner was enforcing the terms of contracts while withholding payments that led to problems for the contractor (Contractor Loses Case Over Payment Delays 2000).

In another case, the Supreme Court of New Hampshire awarded a bidder monetary damages after it determined that a municipality had not awarded the bidder a contract even though it was low bidder. The court ruled that the winning bidder had received special treatment during the bidding process. The trial court initially ruled for the city. The state appellate court reversed the ruling, which was upheld by the State Supreme Court (Cities Beware: Bidders Must be Treated Equally 1994).

A decision by the Massachusetts trial court ruled that a contractor must provide timely notice to the seller of an alleged breach of contract. The federal law cited in the ruling was the Uniform Commercial Code. The trial court's ruling was upheld by the appellate court (Contractor Must Notify Supplier or Get the Goods 1994).

In yet another case, a trial court in Nebraska ruled that an engineering firm was not liable for negligence or breach of contract to a project owner to whom it had no direct contractual relationship. In this case, the owner learned that it had to replace piping in the air-conditioning system because the well water running through it was corrosive. The owner sued the firm that had designed the system under a subcontract to the engineering firm, which had the primary contract with the owner. The trial court ruled that the design firm was not liable because it had no contract with the owner. The appellate court upheld the ruling (Third Party Not Liable for Negligence 1994).

## *Construction Law*

Broadly defined, construction law covers issues that occur during the construction process, including construction documents, performance and design specifications, competitive bidding, contracts, bid bonds, payment bonds, and performance bonds. Other construction business issues may also be included and relate to issues of liability (such as injuries and failures) during the construction process. For example, the Hyatt Hotel failure in the 1970s, (the elevated walkway fell and caused a number of deaths), was the biggest of recent years and resulted in a class action suit. Control of construction with instruments such as codes and standards is part of construction law. Issues such as scheduling, communications, subsoil conditions, suspension of work, sale of material such as ready-mix concrete, delays, and changes are involved.

## *Business Law*

Business law is of great interest to civil engineers. Some of the issues in business law include business organization, regulatory rules, bankruptcy, antitrust, and labor laws. Labor laws, including HR issues, are of increasing concern because of the attention to recent cases through various channels available to employees. Laws governing professional practice as a specialized regulatory issue, are, of course, of interest, as are laws regulating design and construction.

As an example of an award under labor law, construction workers were awarded double wages under a Missouri law that requires general contractors and subcontractors to pay prevailing wages on public contracts. State law gives workers the right to sue for double wages if they have been shortchanged. The award was made by a Missouri trial court and was upheld by the Missouri Court of Appeals (Prevailing Wage Law Takes Double Bite 2000).

Another common business situation is the apportionment of risk, specifically involving insurance companies. In this case, comprehensive-general-liability insurance was ruled to extend to the work of some subcontractors, according to the Supreme Court of Nevada. A developer had built and sold an apartment complex, which began to fall apart due to foundation problems. The insurance company sought to have the damage excluded from coverage, arguing that the building was a product, rather than a service. The court ruled that the site preparation design was a service and was covered. In this case, the trial court's ruling was overturned by the State Supreme Court (Some Subs Are Covered Under Liability Insurance 1993).

## *Tort Law*

Tort law refers to the law of wrongdoing; that is, when one party alleges that another did something wrong against it. Negligence and liability are two of the common issues taken up under tort law. Torts are the legal system's response to injury and provide a means of compensating injured parties. The basic principle behind tort law is that if a person engages in wrongful conduct that is injurious to someone else, then the wrongdoer should compensate the innocent injured party. Reasons for tort law include the following: It benefits society when wrongdoers are forced to pay victims; it deters wasteful injuries and accidents; it deters individuals from behaving in socially wasteful ways; it is economical on the basis of benefit–cost analysis; wrongs should be punished; and it appeals to sense of community justice.

Tort law is primarily a common-law system and is more concerned with reasoning than with rules. Although it is very broad, tort law is encroached on by statutory and regulatory limits. There is an increasing desire to limit tort law—including liability but especially the large damages sometimes awarded by courts (University of Buffalo Law Outline Exchange 1999).

As an example of a liability case, a state trial court ruled that a worker had no claim against the maker of a trailer because he could not prove that the trailer manufacturer was liable for a defective part that led to his injury. The state trial court's verdict was upheld by the State of Virginia Supreme Court (Worker Can't Pin Injury on Trailer 2000).

## Property Law

Property law refers to defining interests in real property, such as real estate, water rights, or personal property, whether tangible (e.g., land, vehicles) or intangible (e.g., stocks and bonds). Property law includes intellectual property and software. A variation of property law deals with intellectual property, such as books, articles, patents, and now, software. This category is becoming much more visible with the Internet and the ease of pirating things that others wrote or produced.

## Family Law

Family law refers to marriage, divorce, child custody, and issues that commonly arise within families. Although this is a large category of civil law, it does not affect the practice of engineering any more than it does other professions and industries.

## Infrastructure Law

Categories of infrastructure law operate both independently, with their own industries, and together, in arenas such as environmental protection and public works management. Each category has its own legal structure.

## Transportation Law

Transportation consists of a complex cluster of areas (highway, rail, transit, air, and water transport), each of which has an array of laws that govern it. Given the large expenditures on roads and bridges, it is likely that law engineers will encounter the Transportation Equity Act (TEA-21) most frequently. The Transportation Equity Act is the umbrella law that provided national funding and guidance for surface transportation in the early part of the century. It provides for financing for projects in the states and for some allocations to mass transit.

There are many related laws, such as the Transportation Infrastructure Finance and Innovation Act, laws that govern speed and safety on roads, and the many laws that govern commerce in transportation. Transportation is a vast area of law that dramatically affects the planning of infrastructure systems. It is interesting to note that the University of Denver College of Law publishes a journal devoted to transportation law.

## Telecommunication Law

Like transportation, telecommunication law is diverse. Telecommunication occurs primarily within the private sector, and laws focus on regulatory issues. For example, breaking up the telephone monopoly in the 1980s was a major legal event in telecommunications. Deregulation continues to occur, especially with the mergers of computing and telecommunications companies.

## *Energy Law*

Energy law is also largely regulatory in nature and deals mainly with electricity and natural gas. The Federal Power Act is one of the main pieces of organizing legislation that deals with, among other issues, hydroelectricity. It requires that utilities give equal consideration to environmental issues. The Electric Consumers Protection Act requires that a project be best adapted to a comprehensive plan for a waterway and requires license conditions to "equitably protect, mitigate damages to, and enhance fish and wildlife, including related spawning grounds and habitat." The Federal Energy Regulatory Commission has responsibility for relicensing of nonfederal hydroelectric power projects.

## *Built Environment Law*

Laws governing the built environment deal with a diverse set of issues ranging from local land use and building codes to major laws setting up such agencies as the Federal Housing Administration, which was aimed at alleviating residential and public housing problems. Health and safety are big regulatory arenas for the built environment, with laws such as the Occupational Health and Safety Act. Civil engineers should become familiar with the range of codes and standards and the promulgating organizations, many of which deal with the built environment.

## *Environmental and Water Law*

Environmental law affects civil engineers a great deal because it has a dual effect: It holds up projects and adds to procedural delays while at the same time creating work for civil engineers (and attorneys) to steer the projects through the resulting mazes of procedural hurdles. Categories of water and environmental law include the following:

- Water rights and allocation
- Interstate compacts
- Water quality
- Endangered species
- Drainage and flooding
- Solid wastes
- Hazardous wastes (including Superfund and toxic substances)
- Pesticides and agricultural environmental issues

- National Environmental Policy Act
- Air quality
- Land use and zoning

## *Working with Attorneys and Expert Witness Work*

Some civil engineers practice expert witness work and are retained by attorneys or clients who need a situation or problem investigated in order to bring to the court an expert opinion about the dispute. For example, someone damaged by a flood may seek to blame a public agency for the damage. A case involving one of the authors (Grigg) occurred when the damaged party, a navigation company, alleged that the Corps of Engineers had operated spillway gates in a way to exacerbate flows and cause a barge to sink. The expert was to analyze the flood sequence, the way the Corps operated the gates, and whether the incident could have been prevented.

Michael D. Bradley wrote a special monograph to guide scientists and engineers serving as expert witnesses (Bradley 1983). He explained the process of civil litigation, how experts work on expert cases, experts and evidence, how courts handle technical information, and tips for experts. There are many surprises for experts serving in court. For example, an expert may think his or her opinion in an area of expertise is without question, but the other side will have their own experts, with independent opinions.

## *Regulatory Law*

Civil engineers may encounter administrative law in the form of rules developed by administrative agencies. These form regulatory law, which is aimed at controlling activities to protect the public interest where private markets do not. For example, regulation in the water industry deals with health and safety, water quality, fish and wildlife, quantity allocation, finance, and service quality. Environmental regulation is a big issue in the United States. Regulatory controls protect the environment; however, controls can be so tight and arbitrary that resentment and political backlash occur. As a result, some are searching for better models than the current statute-by-statute approach to regulation. Ecosystem protection through more comprehensive approaches may be one answer.

The regulatory dilemma faced by water agency and industrial dischargers is one example of regulatory law. The public's interest in water quality is reflected in pressure from both the volunteer side (environmental groups) and the elected side (Congress and state officials). The results eventually come to bear in the form of permits, monitoring, and enforcement of regulations. The press is an important factor, and courts apply pressure, sometimes in response to public opinion. All branches of government get involved in regulatory issues: executive (agencies), legislative (Congress and state legislatures), and judicial (courts). The media gets involved with and influences public opinion on regulatory issues.

Many civil engineers work in regulatory organizations, with their work organized by such laws as the Safe Drinking Water Act; Clean Water Act; Federal Power Act; National Environmental Policy Act; Occupational, Safety, and Health Act; and flood plain regulations. In the western United States, state engineer offices are regulators in the sense that they control the diversion of water from streams and wells. Eastern states are increasing their activity in this area. State natural resource departments with dam safety missions regulate aspects of safety. Similar functions have been developed in the eastern United States. State public service commissions regulate costs of service for utilities. Electric, gas, and telecommunications utilities have their rate decisions made public, and comparisons of costs are easier for the public to make.

## *Law Governing the Professional Practice of Engineering*

Law governing the professional practice of engineering is one of the categories of law left by the federal government to the states. In Colorado, for example, this law is included in the *Colorado Revised Statutes*, Title 12, Professions and Occupations. Article 25 covers engineers and surveyors, and Part 1 of the article covers engineers. Some of the provisions of the article show the goals and controls of the law, which are typical of those of other states (Colorado State Board 1998). The first section in state statute (12-25-101) declares that to:

> safeguard life, health, and property and to promote the public welfare, the practice of engineering is declared to be subject to regulation in the public interest. It shall be deemed that the right to engage in the practice of engineering is a privilege granted by the state through the state board of registration for professional engineers and professional land surveyors, created in section 12-25-106; that the profession involves personal skill and presupposes a period of intensive preparation, internship, due examination, and admission; and that a professional engineer's license is solely such professional engineer's own and is nontransferable.

The remainder of the article in the statute provides detail on the practice of engineering and how it is regulated and contains material that engineers must know to comply with regulations governing their practice. It defines terms such as *engineering practice,* discusses forms of organizations in which engineers are permitted to practice (e.g., individuals, partnerships, corporations), sets up penalties for violations, sets up the state board of registration, provides for disciplinary actions, sets licensing requirements, and lists requirements to become a registered professional engineer.

## Final Word on Law for Civil Engineers

Civil engineers will encounter many aspects of the law, and this chapter has only outlined these issues. Experts advise engineers to know state and local regulations and learn when federal provisions apply. Further, if there is an administrative code that applies to a project or situation, engineers should learn about it.

Attorneys can often be used productively before a problem arises, and it can be advantageous not to wait until it is too late to consult an attorney. Once a problem is created, it is time consuming and expensive to resolve. When conflicts arise, it is always advisable to try mediation or alternative dispute resolution first, if possible.

# 12 Professional Practice and Ethics

## Introduction

This book has discussed important issues of civil engineering professional practice, including heritage, consequences, communication, critical thinking, work principles, design, management, government, law, and economics. These issues, together with technical skills, prepare civil engineers for most areas of work. One issues remains—ethics. This chapter brings together all the issues discussed in earlier chapters on technical skills and provides a perspective on professional practice and ethics.

## Professional Status

Civil engineers are subject to the ethics of society. Such ethics are often expressed in laws and customs as well as in the ethics codes of the civil engineering profession. Laws govern such ethical situations as taking bribes. Society's ethics tell us to be truthful and treat people fairly, but professional ethics take our codes of behavior one step further to cover specific situations that are common to civil engineering work.

What does it mean to be a professional? The literature on sociology and labor economics sheds light on this question and can help us interpret trends. Studies from the 1960s showed that professionalism is a branch of organizational behavior. Robert Perrucci and Joel E. Gerstl, researchers in the field of professional ethics, reported on the landmark study in this area (Perrucci and Gerstl 1969). They found that this period was a time when a number of professions in the United States were undergoing self-examination, and many new and emerging issues needed to be settled. Rapid change had occurred after World War II, and issues of professional responsibility (such as whether an engineer's main task was to work in the public interest rather than for private gain) came under scrutiny.

Professions can be understood in the context of the division of labor, within which tasks are broken into component activities, resulting in end-products that require specialized skills in the labor force. Engineers are generally thought to design and build things, although they also perform many other tasks.

Phases of professionalization are recruitment, training, socialization, membership, improvement, and advancement. Such phases (or professional steps) are similar to the novice, journeyman, and master rankings in guilds. An engineer can progress through the steps of new graduate—passing the Fundamentals of Engineering examination; becoming registered; and taking on the status of a senior, experienced engineer with many years of accomplishment. Professions and crafts descended from guilds, ecclesiastic orders, and universities; and there remains an inherent pecking order associated with them. The older professions of medicine, law, and clergy are generally considered "free" professions (i.e., not tied to government or organizations). Today, however, if a medical doctor works for a health maintenance organization, it may be debatable whether that doctor is a free professional.

Some professions take longer to train their ranks than do others. Engineers graduate in 4 years, whereas doctors and attorneys require additional graduate work. Often, how seriously a profession is taken depends to some extent on the price of entry paid in years of education. Some professions are for life (e.g., doctors and lawyers), whereas others are only a stage in life. It is uncommon for doctors and lawyers to change professions. Engineers, however, are often expected to move into management. Engineering needs a career-long technical track, which offers rewards to those remaining in the technical track; lacking this, it may be difficult to remain an engineer for a full career.

Professions have authority in certain areas. Professionals in one discipline obtain the right to practice in an area closed to others. That is especially true of the medical and law professions, where stringent licensing is enforced. One part of engineering work (requiring that a professional seal be used) is closed. The remainder of the profession is not closed, and there is encroachment from other fields. For example, some engineering work today is done by surveyors, environmental health workers, and landscape architects.

Professionals become professionals by gaining access to a valuable and locked-up body of knowledge. Professions teach both skills and norms. We can see this, for example, in the fact that more civil engineers than other types of engineers become registered. Professions provide standards of judgment in social life and an element of integration in complex society.

There can often be conflict between professionals in organizations and in individual practice. If a professional engineer working in a large organization is asked to do something that is not in the public interest, that engineer may have little recourse. Professions have prestige, and this in the main reasons people remain within a profession. The issue of prestige will continue to be one of the factors during the twenty-first century that determines how civil engineers respond to changes in the character of engineering work and in the engineering profession.

## Engineering as a Profession

Engineering is a relatively new profession, having been organized only for about 200 years. Until the twentieth century, there were few engineering practitioners. Today, there is a proliferation of occupations claiming professional status. Engineers emerged with rapid growth: Where there were 850,000 engineers in 1960, there was less than half that number in 1940. Technologic trends have been the primary impetus for this increase. Although engineering has numerous practitioners, the profession has a great deal of fragmentation. For example, engineering can be divided into the following groups: electrical, mechanical, civil, chemical, and computer. Electrical engineering claims the most numbers. It is interesting to note that civil engineering claims the largest number of sole practitioners (Grigg 2000).

Engineers often lack shared values, resulting in the profession sometimes being referred to as profession without community. Some engineers also have difficulty in cooperating with the various professional associations (e.g., American Association of Engineering Societies). Despite the fact that engineers as a group have difficulty in becoming unified, the ASCE considers civil engineering a profession. In 1963 the ASCE Board adopted the following definition of the profession (ASCE 1994):

> A profession is the pursuit of a learned art in the spirit of public service. ... A profession is a calling in which special knowledge and skill are used in a distinctly intellectual plane in the service of mankind, and in which the successful expression of creative ability and application of professional knowledge are the primary rewards. There is implied the application of the highest standards of excellence in the educational fields prerequisite to the calling, in the performance of services, and in the ethical conduct of its

members. Also implied is the conscious recognition of the profession's obligation to society to advance its standards and to prescribe the conduct of its members.

The focus within the ASCE on professional issues shows awareness of the importance of being a profession. Issues include standards of practice (ethical, business, legal, and administrative) relating to public, private, industrial, educational, and construction engineering. The needs of engineers employed in public agencies, industrial concerns, construction organizations, consulting firms, and educational institutions are also important and have been addressed by the ASCE and throughout this book.

As civil engineering got organized, the profession focused on its purpose. Originally, civil engineers built things, and the ASCE's first code of ethics in 1914 had six articles suggesting that the main focus at that time was private practice. P. Arne Vesiland, a Duke University professor who often writes about professional issues, paraphrased the articles as "do not take bribes, do not speak poorly of colleagues, do not steal work, do not underbid a colleague, do not embarrass a colleague, and do not advertise" (Vesiland 1995).

In 1950 the ASCE adopted the Canon of Ethics prepared by the Engineers Joint Council, which recognized for the first time that the primary goal of civil engineering was to serve the public good. This remained until 1962 when the board changed the preamble to recognize the public interest nature of civil engineering work. In 1976 ASCE abandoned its original code and adopted the Engineer's Council for Professional Development (ECPD) code, which emphasizes public welfare in stating that the engineer's primary responsibility is public health, safety, and welfare. The debate over an environmental article continues today, showing further concern over public issues.

Today, civil engineers in most organizations are encouraged to obtain their professional engineering license. Although it is possible to work as a civil engineer without having a professional engineering license (this is true in many public engineering organizations and even in universities), the license is important because it represents an agreement by the engineer to honor the code of ethics of the profession. In that sense it makes the civil engineer a member of a community bound by a standard of conduct that members of the community agree to honor.

There are excellent resources to support engineers in improving their understanding and application of professional ethics. One such resource is the Online Ethics Center for Engineering and Science. The center provides a variety of information, including an ethics helpline to provide advice on ethical concerns, engineering case studies, essays on ethics, educational resources, and codes of ethics from professional societies such as the ASCE (Online Ethics Center for Engineering and Science 2001).

Chapter 4 discussed the background and growth of civil engineering as an occupational category. With a work force of about 1.5 million, the engineering profession is second in size only to teachers. Civil engineering, at about 200,000 workers, is third behind electrical and mechanical engineering. Aggregation of the work force and economic statistics hide unique characteristics of civil engineering work caused by the concentration on consulting and state and local government. More than 80% of civil engineers work either for consultants or government. Global production of new engineers has passed the one million per year mark, with production in the United States comprising about 12% of the total. These indicators reveal that within the larger civil engineering profession, there are different views of how practice should be conducted and how civil engineers should approach issues such as working for organizations versus being in private practice (Grigg 2000).

# Professional Societies

Some aspects of professions came out of the guilds. Today the closest organization there is to a guild is the professional society. Engineering professional societies do many things, including organizing the work of the profession, controlling access, guiding the education institutions, setting standards for work, and advocating for recognition and prestige.

One indicator of the situation in the ASCE is its *Strategic Plan*. After five years of work, beginning in 1993, ASCE published its *Strategic Plan* with six categories of driving forces (ASCE 1998b):

- A changing marketplace is creating design–build, turnkey projects, privatization, joint ventures, and similar developments that affect the way civil engineers do business.
- Specialization within the profession threatens to splinter off engineers into specialty organizations apart from ASCE.
- Encroachment by professionals in related fields (e.g., hydrologists, geologists, planners) and paraprofessionals (e.g., technologists, technicians) is impacting traditional civil engineering positions.
- Undercompensation relative to the importance and level of liability is a fact of life within the civil engineering profession.
- Loss of prominence among government and private industry policymakers hampers professional progress.
- A bureaucratic society, chasing too many programs with too few resources, has limited its ability to make a serious impact in high-priority areas.

Popular terminology (soundbites) for these trends includes industry dynamics, specialization and splintering, encroachment, undercompensation, loss of influence, and a bureaucratic professional society.

## Engineering Education

There is a strong connection between professionalization and professional schools. To engage in continuous improvement of civil engineering education, we must find out the needs of industry, recruit students, organize and deliver the curriculum, and assess how well we are doing. This is what ABET requires. In the 1990s, civil engineering enrollments declined. The phenomenon is cyclic and civil engineering enrollments have not been as volatile as have enrollments in other fields, but the changes still deserve comment. David E. Daniel, Head of the CE Department at the University of Illinois, probed this question, suggesting the reason is that job opportunities and salaries are stronger in other fields. Daniel pointed to the level of professionalism being of concern, citing data that showed that young engineers are not allowed to attend conferences. He also thought that civil engineers are too narrow, noting that engineering in the service sector requires a broad view. He also suggested that civil engineers have built barriers around the profession using licensing and procurement procedures. In this way, he notes, civil engineering limits itself. Daniel concluded that skills needed today are different than those needed 30 years ago, and civil engineers must expand their horizons (Daniel 1999).

## Defining Characteristics of Civil Engineering

Data about civil engineering suggest hypotheses about civil engineering professional practice, including the following:

- Civil engineering is a generally stable profession.
- Civil engineering requires more management skills than do most other engineering disciplines (although management is largely ignored in educational programs).
- Civil engineering is similar to other engineering disciplines in technical content.
- Civil engineering is more splintered than are other engineering disciplines.
- Civil engineering splintering and competition are caused by the growing complexity of knowledge and by self-interest of individuals and their associations.
- Civil engineering is driven by government more than are other professions.
- Civil engineering has a greater social component than do the other disciplines.
- Civil engineering is not growing as quickly as are other engineering disciplines.

- Civil engineering is not as technical-commercial as are other engineering disciplines.
- Civil engineering has a more professional influence on construction and infrastructure industries than does any other profession.
- Civil engineering has more influence on the environment than do other engineering disciplines and professions, including professional, economic, and policy/legal influence.

Chapter 2 detailed that the future will be different for civil engineers. Although the future cannot be accurately predicted, one can consider different possibilities. Defining characteristics for civil engineering can serve as a guide to the future. Some important issues discussed by Neal Peirce, a syndicated columnist, will affect the modes of professional practice (Table 12-1).

**Table 12-1** The impact of Peirce's predictions on civil engineering.

| Peirce's Predictions (1999) | Potential Impact On Civil Engineering |
|---|---|
| Big global population increase—from 6 billion today to 10 billion in 2050 | This will drive tremendous needs for new facilities and environmental controls. |
| More sprawl, with traffic congestion, wasted land, environmental quandaries, and isolation of poor in urban centers | Civil engineers should lead in the search for solutions to these difficult problems. |
| Geographic information systems and other intelligent transportation applications will grow | Civil engineering must change to adapt new technologies. |
| A tremendous surge in business over the Internet | Civil engineering work will be shared over the Internet between offices located around the world. |
| American's high consumption rate, equivalent to 180 Bangladeshans, comes under scrutiny | Civil engineers continue to lead the search for sustainable engineering technologies. |
| Changes in local government, the biggest providers of infrastructure | Civil engineers will need to know how local government is changing and facilitate the changes with improved practices. |
| Economic disparities between parts of cities and suburbia | Civil engineering work involves social as well as technical issues and should offer methods to overcome disparities. |
| Key words will be cooperation, collaboration, communication, networking, alliances, joint ventures, partnerships | Civil engineering work will change to embrace these concepts. |
| No issue is the sole province of one government, all are intergovernmental, with more coordination needed | Through their work as consultants and managers, civil engineers will serve as coordinators. |
| A search for new ways to tax Internet transactions | This will have little effect on civil engineering. |

# Engineering Ethics

Obviously, professionals are subject to the same ethical responsibilities as are other citizens as well as to additional responsibilities that deal with specific issues within their work areas. The ethics of medical doctors, ministers, lawyers, and stockbrokers are often in the news. Civil engineers are subject to several codes of ethics that have been developed over the years.

Although this section does not detail ethical situations, the Online Ethics Center for Engineering and Science (2001) provides 36 discussion items based on cases considered by the Board of Ethical Review of the National Society of Professional Engineers. These fall into five categories: public safety and welfare, conflicting interests and conflict of interest, ethical engineering/fair trade practices, international engineering ethics, and research ethics. Titles of selected cases are given below to illustrate the range of topics that can be considered:

- Suspected Hazardous Waste
- Code Violations with Safety Implications
- The Whistle-Blowing City Engineer
- Engineer's Dispute with Client over Design
- The Right of Engineers to Have a Right to Protest Shoddy Work and Cost Overruns
- Changes in Statement of Qualifications for a Public Project
- Knowledge of Damaging Information
- Engineering Student Serving as Consultant to the University
- Conflict of Interest in a Feasibility Study
- Accepting a Complimentary Seminar Registration
- Engineer's Disclosure of Potential Conflict of Interest
- Related Work Done for a Private Party Following Public Employment
- Signing Off on Drawings
- Intellectual Property of Engineers in Private Practice
- An Engineer's Agreement with Two Competing Firms for the Same Contract
- Competition from Former Employees
- Maintaining Professional Standards: Writing a Letter of Recommendation
- The Use of Work from an Unpaid Consultation
- Promotional Letter Emphasizing Negative Attributes of Other Firms
- Gifts to Foreign Officials
- Improper Credit Given for Research Data

These are valuable topics for engineers to consider and to use for discussion purposes in the classroom. To provide a background for further reading and discussion, ASCE's Code of Ethics is included in this book as an appendix.

## Importance for Professions

If the civil engineering profession is to add significantly to solutions to these issues, it will have to emphasize professionalism more in education. We must better explain infrastructure and environmental industries, including government, to the public. Civil engineers need to focus on decision making in the public arena. Technical education must continue, but other components must be added, especially management topics.

Civil engineering's heritage includes great structures, complex transportation systems, environmental controls to restore the balance of nature, and complex components and systems of infrastructure. Today, the evolving nature of civil engineering work will require more sophisticated management systems for public infrastructure and environmental systems.

G. Wayne Clough's view of the future of civil engineering (see Chapter 1) noted that the growing world population and human tendency to put off infrastructure and environmental improvements will define civil engineering work in the future. In order to apply new management solutions and technologies, civil engineers in the next millennium must be educated differently than in the past (Clough 2000). Civil engineering as a profession must also work to attract and keep its share of the best and brightest amid a rapidly changing and diversifying work force.

# Appendix: ASCE Code of Ethics

As adopted September 2, 1914 and most recently amended November 10, 1996.

## Fundamental Principles*

Engineers uphold and advance the integrity, honor and dignity of the engineering profession by:

1. using their knowledge and skill for the enhancement of human welfare and the environment;
2. being honest and impartial and serving with fidelity the public, their employers and clients;
3. striving to increase the competence and prestige of the engineering profession; and
4. supporting the professional and technical societies of their disciplines.

---

*The American Society of Civil Engineers adopted THE FUNDAMENTAL PRINCIPLES of the ABET Code of Ethics of Engineers as accepted by the Accreditation Board for Engineering and Technology, Inc. (ABET). (By ASCE Board of Direction action April 12-14, 1975)

## Fundamental Canons

1. Engineers shall hold paramount the safety, health and welfare of the public and shall strive to comply with the principles of sustainable development* in the performance of their professional duties.
2. Engineers shall perform services only in areas of their competence.
3. Engineers shall issue public statements only in an objective and truthful manner.
4. Engineers shall act in professional matters for each employer or client as faithful agents or trustees, and shall avoid conflicts of interest.
5. Engineers shall build their professional reputation on the merit of their services and shall not compete unfairly with others.
6. Engineers shall act in such a manner as to uphold and enhance the honor, integrity, and dignity of the engineering profession.
7. Engineers shall continue their professional development throughout their careers, and shall provide opportunities for the professional development of those engineers under their supervision.

## Guidelines to Practice under the Fundamental Canons of Ethics

CANON 1. Engineers shall hold paramount the safety, health and welfare of the public and shall strive to comply with the principles of sustainable development in the performance of their professional duties.

a. Engineers shall recognize that the lives, safety, health and welfare of the general public are dependent upon engineering judgments, decisions and practices incorporated into structures, machines, products, processes and devices.
b. Engineers shall approve or seal only those design documents, reviewed or prepared by them, which are determined to be safe for public health and welfare in conformity with accepted engineering standards.
c. Engineers whose professional judgment is overruled under circumstances where the safety, health and welfare of the public are endan-

---

*In November 1996, the ASCE Board of Direction adopted the following definition of Sustainable Development: "Sustainable Development is the challenge of meeting human needs for natural resources, industrial products, energy, food, transportation, shelter, and effective waste management while conserving and protecting environmental quality and the natural resource base essential for future development."

gered, or the principles of sustainable development ignored, shall inform their clients or employers of the possible consequences.

d. Engineers who have knowledge or reason to believe that another person or firm may be in violation of any of the provisions of Canon 1 shall present such information to the proper authority in writing and shall cooperate with the proper authority in furnishing such further information or assistance as may be required.

e. Engineers should seek opportunities to be of constructive service in civic affairs and work for the advancement of the safety, health and well-being of their communities, and the protection of the environment through the practice of sustainable development.

f. Engineers should be committed to improving the environment by adherence to the principles of sustainable development so as to enhance the quality of life of the general public.

CANON 2. Engineers shall perform services only in areas of their competence.

a. Engineers shall undertake to perform engineering assignments only when qualified by education or experience in the technical field of engineering involved.

b. Engineers may accept an assignment requiring education or experience outside of their own fields of competence, provided their services are restricted to those phases of the project in which they are qualified. All other phases of such project shall be performed by qualified associates, consultants, or employees.

c. Engineers shall not affix their signatures or seals to any engineering plan or document dealing with subject matter in which they lack competence by virtue of education or experience or to any such plan or document not reviewed or prepared under their supervisory control.

CANON 3. Engineers shall issue public statements only in an objective and truthful manner.

a. Engineers should endeavor to extend the public knowledge of engineering and sustainable development, and shall not participate in the dissemination of untrue, unfair or exaggerated statements regarding engineering.

b. Engineers shall be objective and truthful in professional reports, statements, or testimony. They shall include all relevant and pertinent information in such reports, statements, or testimony.

c. Engineers, when serving as expert witnesses, shall express an engineering opinion only when it is founded upon adequate knowledge of the facts, upon a background of technical competence, and upon honest conviction.

d. Engineers shall issue no statements, criticisms, or arguments on engineering matters which are inspired or paid for by interested parties, unless they indicate on whose behalf the statements are made.

e. Engineers shall be dignified and modest in explaining their work and merit, and will avoid any act tending to promote their own interests at the expense of the integrity, honor and dignity of the profession.

CANON 4. Engineers shall act in professional matters for each employer or client as faithful agents or trustees, and shall avoid conflicts of interest.

a. Engineers shall avoid all known or potential conflicts of interest with their employers or clients and shall promptly inform their employers or clients of any business association, interests, or circumstances which could influence their judgment or the quality of their services.

b. Engineers shall not accept compensation from more than one party for services on the same project, or for services pertaining to the same project, unless the circumstances are fully disclosed to and agreed to, by all interested parties.

c. Engineers shall not solicit or accept gratuities, directly or indirectly, from contractors, their agents, or other parties dealing with their clients or employers in connection with work for which they are responsible.

d. Engineers in public service as members, advisors, or employees of a governmental body or department shall not participate in considerations or actions with respect to services solicited or provided by them or their organization in private or public engineering practice.

e. Engineers shall advise their employers or clients when, as a result of their studies, they believe a project will not be successful.

f. Engineers shall not use confidential information coming to them in the course of their assignments as a means of making personal profit if such action is adverse to the interests of their clients, employers or the public.

g. Engineers shall not accept professional employment outside of their regular work or interest without the knowledge of their employers.

CANON 5. Engineers shall build their professional reputation on the merit of their services and shall not compete unfairly with others.

a. Engineers shall not give, solicit or receive either directly or indirectly, any political contribution, gratuity, or unlawful consideration in order to secure work, exclusive of securing salaried positions through employment agencies.

b. Engineers should negotiate contracts for professional services fairly and on the basis of demonstrated competence and qualifications for the type of professional service required.

c. Engineers may request, propose or accept professional commissions on a contingent basis only under circumstances in which their professional judgments would not be compromised.

d. Engineers shall not falsify or permit misrepresentation of their academic or professional qualifications or experience.

e. Engineers shall give proper credit for engineering work to those to whom credit is due, and shall recognize the proprietary interests of others. Whenever possible, they shall name the person or persons who may be responsible for designs, inventions, writings or other accomplishments.

f. Engineers may advertise professional services in a way that does not contain misleading language or is in any other manner derogatory to the dignity of the profession. Examples of permissible advertising are as follows:
- Professional cards in recognized, dignified publications, and listings in rosters or directories published by responsible organizations, provided that the cards or listings are consistent in size and content and are in a section of the publication regularly devoted to such professional cards.
- Brochures which factually describe experience, facilities, personnel and capacity to render service, providing they are not misleading with respect to the engineer's participation in projects described.
- Display advertising in recognized dignified business and professional publications, providing it is factual and is not misleading with respect to the engineer's extent of participation in projects described.
- A statement of the engineers' names or the name of the firm and statement of the type of service posted on projects for which they render services.
- Preparation or authorization of descriptive articles for the lay or technical press, which are factual and dignified. Such articles shall not imply anything more than direct participation in the project described.
- Permission by engineers for their names to be used in commercial advertisements, such as may be published by contractors, material suppliers, etc., only by means of a modest, dignified notation acknowledging the engineers' participation in the project described. Such permission shall not include public endorsement of proprietary products.

g. Engineers shall not maliciously or falsely, directly or indirectly, injure the professional reputation, prospects, practice or employment of another engineer or indiscriminately criticize another's work.

h. Engineers shall not use equipment, supplies, laboratory or office facilities of their employers to carry on outside private practice without the consent of their employers.

CANON 6. Engineers shall act in such a manner as to uphold and enhance the honor, integrity, and dignity of the engineering profession.

a. Engineers shall not knowingly act in a manner which will be derogatory to the honor, integrity, or dignity of the engineering profession or knowingly engage in business or professional practices of a fraudulent, dishonest or unethical nature.

CANON 7. Engineers shall continue their professional development throughout their careers, and shall provide opportunities for the professional development of those engineers under their supervision.

a. Engineers should keep current in their specialty fields by engaging in professional practice, participating in continuing education courses, reading in the technical literature, and attending professional meetings and seminars.

b. Engineers should encourage their engineering employees to become registered at the earliest possible date.

c. Engineers should encourage engineering employees to attend and present papers at professional and technical society meetings.

d. Engineers shall uphold the principle of mutually satisfying relationships between employers and employees with respect to terms of employment including professional grade descriptions, salary ranges, and fringe benefits.

# References

Accreditation Board for Engineering and Technology Inc. 1999. Engineering Criteria 2000. Criteria for Accrediting Engineering Programs during the 1999–2000 Accreditation Cycle. Available from www.abet.org; INTERNET.

American Academy of Environmental Engineers. 1999. Available from www.enviro-engrs.org/history.htm; INTERNET.

American Association of Engineering Societies. 1995. Washington, D.C.: Annual Report of the Engineering Workforce Commission.

American Consulting Engineers Council. 1999. About the ACEC. Available from www.asce.org; INTERNET.

American Consulting Engineers Council of Colorado. 1999. Top ten crises faced by design firms. *Monthly News*, 2:5.

American Council for Construction Education. 1999. Available from www.calpoly.edu/~cm/acce; INTERNET.

American Society of Civil Engineers. 1990. *Quality in the Constructed Project: A Guide for Owners, Designers and Constructors. ASCE Manuals and Reports on Engineering Practice No. 73*. Reston, Va: ASCE.

American Society of Civil Engineers. 1994. Committee on Professional Practice Div and Committee on Standards of Practice. COSP. p 188.

American Society of Civil Engineers. 1995. 1995 Civil Engineering Education Conference, Denver, Colorado, June 8–11, 1995.

American Society of Civil Engineers. 1996. *Consulting Engineering: A Guide for the Engagement of Engineering Services Manual 45*. Reston, Va: ASCE.

American Society of Civil Engineers. 1997. *Official Register*. Reston, Va: ASCE.

American Society of Civil Engineers. 1998a. Strategic Plan. Available from www.asce.org; INTERNET.

American Society of Civil Engineers. 1998b. What is Civil Engineering? Available from www.eranch.com/whatis.html; INTERNET.

American Society of Civil Engineers. 1999. Geo-Institute and Structural Engineering Institute. Available from www.asce.org; INTERNET.

American Society of Civil Engineers. 2000. ASCE names top achievements of 20th century. *ASCE News*. 25:1–6.

American Water Works Research Foundation. 1995. Project Update, Drinking Water Research, January/February.

Armstrong, Ellis L., ed. 1976. *History of Public Works in the United States.* Chicago: American Public Works Association, Public Works Historical Society.

Augustine, Norman R. 1996. Rebuilding engineering education. *Chronicle of Higher Education*. May 24.

Baetz, B. W., and R. M. Korol. 1995. Evaluating technical alternatives on basis of sustainability. *Journal of Professional Issues in Engineering Education and Practice*. 121:102–107.

Beakley, George C., Donovan L. Evans, and Deloss H. Bowers. 1987. *Careers in Engineering and Technology*. 4th ed. New York: Macmillan.

Bechtel Corporation. 1999. Available from http://www.bechtel.com/aboutbech/timeline.html; INTERNET.

Beer, D., and D. McMurrey. 1999. *A Guide to Writing as an Engineer*. New York: John Wiley & Sons.

Billington, D. P. 1983. *The Tower and the Bridge: The New Art of Structural Engineering*. Princeton, NJ: Princeton University Press.

Bok, Derek. 1996. *The State of the Nation: Government and the Quest for a Better Society*. Cambridge, Mass: Harvard University Press.

Bradley, Michael D. 1983. *The Scientist and Engineer in Court*. Washington, D.C.: American Geophysical Union.

Burnham, William. 1995. *Introduction to the Law and Legal System of the United States*. St. Paul, Minn: West Publishing Company.

Burton, Lawrence, Linda Parker, and William K. LeBold. 1998. US engineering career trends. *ASEE Prism*. 79:18–21.

California Department of Transportation. 1999. Available from http://www.dot.ca.gov/hq/paffairs/ about/100yrs.htm; INTERNET.

Carter, Jimmy. 1996. *Why not the Best: The First Fifty Years*. Fayetteville, Ark: University of Arkanss Press.

Center for Civic Education. 1994. Information available from http://www.civiced.org/stds; INTERNET.

Center for Global Studies. 1993. Towards understanding sustainability. *Woodlands Forum*. 10:1.

Chaffee, J. 1998. *The Thinker's Way: 8 Steps to a Richer Life*. Boston: Little, Brown and Company.

Choate, Pat, and Susan Walter. 1981. *America in Ruins: Beyond the Public Works Pork Barrel*. Washington, D.C.: Council of State Planning Agencies.

Cities beware: Bidders must be treated equally. 1994. *Civil Engineering*. 64:30.

City of Charlotte, North Carolina. 1999. Available from http://www.charmeck.nc.us; INTERNET.

City of Fort Collins, Colorado. 2001. Available from http://www.ci.fort-collins.co.us; INTERNET.

City of San Francisco. 1999. Web page. http://www.ci.sf.ca.us/deptsall.htm.

Clough, G. Wayne. 2000. Civil Engineering in the Next Millenium. Paper read at MIT New Millennium Colloquium, Cambridge, Mass.; March 20–21.

Colorado State Board of Registration for Professional Engineers and Professional Land Surveyors 1998. 73rd Annual Report. Denver.

Committee on Water Resources Planning. 1962. Basic considerations in water resources planning. *Journal of the Hydraulics Division.*

Condreay, Scott, and Timothy McCune. 1984. A summary report of the infrastructure rebuilding management challenge conference. Fort Collins, Colo.: Colorado State University.

Contractor loses case over payment delays. 2000. *ENR.* 244:22.

Contractor must notify supplier or get the goods. 1994. *Civil Engineering.* 64:30.

Covey, Stephen R. 1991. *Principle-Centered Leadership.* New York: Summit Books.

Creech, Bill. 1994. *The Five Pillars of TQM: How to Make Total Quality Management Work for You.* New York: Truman Talley Books/Plume.

Creighton, James L. 1981. *The Public Involvement Manual.* Cambridge, Mass.: Abt Books.

Cross, N. 1994. *Engineering Design Methods: Strategies for Product Design.* Chichester, England: John Wiley & Sons.

Cunningham, J. Barton, and John Farquharson. 1989. Systems problem-solving: Unraveling the "mess." *Management Decision.* 27:1.

Daniel, David E. 1999. On the role of civil engineers in non-traditional areas. Civil and Environmental Engineering Alumni Association. Chicago: University of Illinois, Fall:4–5.

Daniels, Jane Zimmer. 1988. Women in engineering: A program administrator's perspective. *Engineering Education.* 78:766–768.

D'Arcy, J. 1998. *Technically Speaking: A Guide for Communicating Complex Information.* Columbus, Ohio: Battelle Press.

Davies, J. W. 1996. *Communication for Engineering Students.* Essex, England: Longman Group Limited.

The decaying of America. 1982. *Newsweek.* August 2.

Dewey, J. 1933. *How We Think.* Lexington, Mass: D.C. Heath.

Drucker, Peter F. 1954. *The Practice of Management.* New York: Harper & Row.

Drucker, Peter F. 1976. *Management: Tasks, Responsibilities, Practices.* New York: Harper and Row.

Dym, C. L. 1994. *Engineering Design: A Synthesis of Views.* Cambridge, England: Cambridge University Press.

Dym, C. L., and P. Little. 2000. *Engineering Design: A Project-Based Approach.* New York: John Wiley & Sons.

Eisenberg, A. 1992. *Effective Technical Communication.* New York: McGraw-Hill.

Ellis, R.A. 1997. SESTAT: New data on engineers from the National Science Foundation. *Engineers.* 3:2–13.

Environmental and Energy Study Institute Task Force. 1991. Partnership for Sustainable Development: A New US Agenda for International Development and Environmental Security. Washington, D.C.

Fair, Gordon M., J. C. Geyer, and D. A. Okun. 1966. *Water and Wastewater Engineering.* New York: John Wiley & Sons.

Fisher, Roger, and William Ury. 1981. *Getting to Yes: Negotiating Agreement Without Giving In.* New York: Penguin Books.

Fogg, C. Davis. 1994. *Team-Based Strategic Planning: A Complete Guide to Structuring, Facilitating, and Implementing the Process.* New York: AMACOM.

Forrester, Jay W. 1961. *Industrial Dynamics.* Cambridge, Mass: MIT Press.

Forrester, Jay W. 1969. *Urban Dynamics*. Cambridge, Mass: MIT Press.

The fourteen points of Deming. 1988. *Coloradoan*. December 18.

Frontius, Sextus Julius. 1973. *The Two Books on the Water Supply for the City of Rome, AD 97*. Translated by Clemens Herschel. Boston: republished by the New England Water Works Association.

Government Finance Research Center. 1981. *Planning for Clean Water Programs: Role of Financial Analysis*. Washington, D.C.: US Government Printing Office.

Greene, James H. 1984. *Operations Management: Productivity and Profit*. Reston, Va: Reston Publishing Company.

Greenough, Geoff, Thomas Eggum, Ulysses G. Ford, III, Neil S.Grigg, and Ed Sizer. 1999. Public works delivery systems in North America: Private and public approaches, including managed competition. *Journal of Public Works Planning and Management*. 4(1):41–49.

Grigg, Neil S. 1996. *Water Resources Management: Principles, Regulations, and Cases*. New York: McGraw-Hill.

Grigg, Neil S. 1988. *Infrastructure Engineering and Management*. New York: John Wiley & Sons.

Grigg, Neil S. 2000. Demographics and industry employment of the civil engineering workforce. *Journal of Professional Issues in Engineering Education and Practice*. 126(3):1–9.

Grigg, Neil S., C. F. Andersen, J. Z. Daniels, and J. G. A. Levine. 1997. Women in public works: the engineering pipeline. *Public Works Management & Policy*. 2(2):121–128.

Gutfeld, Rose. 1993. Agreement is reached on framework of $465 million plan to save Everglades. *Wall Street Journal*. July 14.

Hatry, Harry P. 1980. Performance measurement principles and techniques: An overview for local government. *Public Productivity Review*. December.

Hatry, Harry P. and George E. Peterson. 1984. Guides to managing urban capital. Washington, D.C.: Urban Institute.

Heilbroner, Robert, and Lester Thurow. 1994. *Economics Explained*. New York: Touchstone Books.

History Channel. 1999. Available from http://www.historychannel.com/skyscrapers/; INTERNET.

Hofferbert, Richard I., and David L. Cingranelli. 1996. Public policy and administration: Comparative policy analysis. In Goodlin, Robert E, and Hans-Dieter Klingemann, eds. *A New Handbook of Political Science*. Oxford, England: Oxford Press.

Holmes, Beatrice Hort. 1972. *A History of Federal Water Resources Programs, 1800–1960*. Washington, D.C.: US Department of Agriculture, Economic Research Service.

Holmes, Beatrice Hort. 1979. *History of Federal Water Resources Programs and Policies, 1961–1970*. Washington, D.C.: US Department of Agriculture, Economics, Statistics and Cooperatives Service. Misc. Publication No. 1379.

Hudson, W. Ronald, Ralph Haas, and Waheed Uddin. 1997. *Infrastructure Management: Design, Construction, Maintenance, Rehabilitation, Renovation*. New York: McGraw-Hill.

Hyman, David N. 1989. *Economics*. Boston: Irwin.

Johnson, C. D., and R. M. Korol. 1995. Incorporating Sustainable Development Principles in the Civil Engineering Curriculum—An Urgent Need. *Fourth World Conference On Engineering Education*. Saint Paul, Minn:197–201.

Johnston, William B., and Arnold E. Packer. 1987. *Workforce 2000: Work and Workers for the Twenty-first Century.* Indianapolis, Ind.: Hudson Institute.

Judy, Richard W., and Carol D'Amico. 1997. *Workforce 2020: Work and Workers in the 21st Century.* Indianapolis, Ind.: Hudson Institute.

Katko, Tapio S. 1997. *Water: Evolution of Water Supply and Sanitation in Finland from the mid-1800s to 2000.* Finnish Water and Waste Water Works Association.

Keller, David N. 1989. *Stone & Webster, 1889–1989, A Century of Integrity and Service.* New York: Stone & Webster.

Kirby, R. S., S.Withington, A.B. Darling, and F. G. Kilgour. 1956. *Engineering in History*. New York: McGraw-Hill.

Koen, B. V. 1985. *Definition of the Engineering Method.* Washington, D.C.: American Society for Engineering Education.

Koeppel, G. 1994. A struggle for water. *Invention and Technology*. 18–30.

Korbitz, William E., ed. 1976. *Urban Public Works Administration*. Washington, D.C.: International City Management Association.

Kottegoda, N.T., and R. Rosso. 1997. *Statistics, Probability, and Reliability for Civil and Environmental Engineers*. New York: McGraw-Hill.

Kraemer, Kenneth L. 1973. *Policy Analysis in Local Government*. Washington, D.C.: International City Management Association.

Krick, E.V. 1969. *Engineering and Engineering Design*. New York: John Wiley & Sons.

Lewis, Tom. 1997. *Divided Highways: Building the Interstate Highways, Transforming American Life.* New York: Viking Penguin.

Lienhard, John H. 1998. The Polytechnic Legacy. Dallas: ASME Management Training Workshop. Available from http://www.uh.edu/engines/asmedall.html; INTERNET.

Life after communism. 1999. *Wall Street Journal*. November 17.

Lincoln, Samuel B. 1960. *Lockwood Greene: The History of An Engineering Business, 1832-1958*. Brattleboro, Vt: The Stephen Greene Press.

Lipman, M. 1988. Critical thinking: what can it be? *Educational Leadership*. 46:38-43.

Lock, Dennis, ed. 1987. *Project Management Handbook*. Aldershot, England: Gower Technical Press.

Love, S.F. 1986. *Planning and Creating Successful Engineering Designs: Managing the Design Process*. Los Angeles: Advanced Professional Development, Inc.

Magretta, Joan, and Nan Stone. 1999. The original Management Guru. *Wall Street Journal*. November 19.

Martin, James L. 1986. In Cristofano, Sam M., and William S. Foster, eds. *Management of Local Public Works*, Washington, D.C.: International City Management Association.

Martin M.W., and R. Schinzinger. 1989. *Ethics in Engineering*. New York: McGraw Hill.

Michaelson, H.B. 1990. *How to Write & Publish Engineering Papers and Reports*. Phoenix, Ariz: Oryx Press.

Mileti, D. S. 1999. *Disasters by Design: A Reassessment of Natural Hazards in the United States.* Washington, D.C.: Joseph Henry Press.

Minkler, John. 1998. Active citizenship: Empowering America's youth. Available from http:www.activecitizenship.org/curriculum.html; INTERNET.

Moak, Lennox L., and Albert M. Hillhouse. 1975. *Concepts and Practices in Local Government Finance,* Chicago: Municipal Finance Officer's Association.

Molof, Alan H., and Carl J. Turkstra, eds. 1984. *Infrastructure: Maintenance and Repair of Public Works.* New York: New York Academy of Sciences.

Moody, Paul E. 1983. *Decision Making: Proven Methods for Better Decisions.* New York: McGraw-Hill.

More civics, please. 1999. *Denver Post.* November 20.

National Academy of Engineering. 1989. Top 10 Achievements: 1964–1989, December 5.

National Academy of Engineering. 2000. Available from www.greatachievements.org; INTERNET.

National Council on Engineering Examiners. 2001. Available from www.ncees.org; INTERNET.

National Science Foundation. 1994. *Civil Infrastructure Systems, An Integrative Research Program, Program Announcement and Guidelines.* Washington, D.C.: NSF.

National Science Foundation. 1995. *Civil Infrastructure Systems: An Integrative Research Program.* Washington, D.C.: NSF.

Nunis, Jr., B. Doyce. 1995. *The St. Francis Dam Disaster Revisited.* Los Angeles: Historical Society of Southern California.

Oakes, William C., Les L. Leone, Craig J. Gunn, John B. Dilworth, Merle C. Potter, Michael F. Young, Heidi A. Diefes, and Ralph E. Flori. 1999. *Engineering Your Future.* Wildwood, Mo: Great Lakes Press.

Oglesby, Clarkson H. 1975. *Highway Engineering.* 3d ed. New York: John Wiley & Sons.

Online Ethics Center for Engineering and Science. 2001. A general guide to the Online Ethics Center. Available from http://onlineethics.org; INTERNET.

O'Keefe, R.M., O. Balci, and E.P. Smith. 1987. Validating expert system performance. *IEEE Expert.* 2:81–90.

Peirce, Neal. 1999. Reform or Fall Victim: 21st Century Challenges for Local Government. Available from www.govtech.net; INTERNET.

Perrucci, Robert, and Joel E. Gerstl. 1969. *Profession Without Community: Engineers in American Society.* New York: Random House.

Petroski, H. 1995. *Engineers of Dreams: Great Bridge Builders and the Spanning of America.* New York: Alfred A. Knopf.

Petroski, Henry. 1996. *Invention by Design: How Engineers Get from Thought to Thing.* Cambridge, Mass: Harvard University Press.

Petroski, H. 1997. *Remaking the World: Adventures in Engineering.* New York: Vintage Books.

Prasuhn, Alan. 1995. Early history of professional registration in United States. *Journal of Professional Issues in Engineering Education and Practice.* 121:25–29.

Prendergast, Candice. 1993. A theory of "Yes Men." *American Economic Review.* September.

President's Council on Sustainable Development. 1994. *Challenges to Natural Resource Management and Protection of the Colorado River Basin*, University of Nevada Las Vegas.

Prevailing wage law takes double bite. 2000. *ENR*. 244:22.

Priscoli, Jerome Delli. 1989. Public involvement, conflict management: Means to EQ and social objectives. *Journal of Water Resources Planning and Management*. 115:31–41.

Putnam, Robert D. 1995. Bowling alone: America's declining social capital. *Journal of Democracy*. 6:65–78.

Richman, Barry M., and Richard N. Farmer. 1975. *Management and Organizations*. New York: Random House.

Rudwick, Bernard H. 1973. *Systems Analysis for Effective Planning*. New York: John Wiley & Sons.

Schreiber, Arthur F., Paul K. Gatons, and Richard B. Clemmer. 1971. *Economics of Urban Problems: An Introduction*. Boston: Houghton Mifflin.

Senge, Peter M. 1990. *The Fifth Discipline: The Art and Practice of the Learning Organization*. New York: Doubleday Currency.

Sheeran, F. Burke. 1976. *Management Essentials for Public Works Administrators*. Chicago: American Public Works Association.

Smith, Craig B. 1999. Program management B.C. *Civil Engineering*. June.

Some subs are covered under liability insurance. 1993. *Civil Engineering*. 63:30.

South Florida Water Management District. 1989. 40th Annual Report, 1988–89, West Palm Beach, Florida.

South Florida Water Management District. 1999. Available from www.sfwmd.gov; INTERNET.

Spillinger, Ralph S. 1999. *Adding Value to the Facility Acquisition Process: Best Practices for Reviewing Facility Designs. Federal Facilities Council Technical Report 139*. Washington, D.C.: National Academy Press.

State of North Carolina. 1999. http://www.state.nc.us

Stone, Donald. 1974. *Professional Education in Public Works/Environmental Engineering and Administration*. Chicago: American Public Works Association and National Association of Schools of Public Affairs & Administration.

Tarr, Joel A. 1984. The evolution of the urban infrastructure in the nineteenth and twentieth centuries. In *Perspectives on Urban Infrastructure*, Royce Hanson, ed. Washington, D.C.: National Academy Press.

Third party not liable for negligence. 1994. *Civil Engineering*. 64:30.

To rebuild America: $2,500,000 job. 1982. *US News and World Report*. September 27.

University of Buffalo Law Outline Exchange. 1999. Available from http://wings.buffalo.edu/law/latis/unub/exchange/; INTERNET.

University of Waterloo Career Development Manual. 1997. Available from www.adm.uwaterloo.ca; INTERNET.

US Army Corps of Engineers. 1984. Waterways experiment station. *The REMR Bulletin*. Vicksburg, Miss: The REMR Research Program.

US Bureau of Reclamation. 1973. *Design of Small Dams*. 2d ed. Washington, D.C.: US Government Printing Office.

US Committee on Transportation and Infrastructure. 2001. A Brief History. Available from http://www.house.gov/transportation/history/history1.htm; INTERNET.

US Department of Commerce, Bureau of Census. 1977. Washington, D.C.:1977 Census of the Service Industry.

US Department of Commerce. 1998. Washington, D.C.: Statistical Abstract of the United States, 118th Edition.

US Federal Government Agencies Directory. 1999. Available from http://www.lib.lsu.edu/gov/exec.html; INTERNET.

US General Accounting Office. 1982. *Effective Planning and Budgeting Practices Can Help Arrest the Nation's Deteriorating Public Infrastructure.* Washington, D.C.: US General Accounting Office.

US House of Representatives. 1999. Available from http://www.house.gov/transportation; INTERNET.

US Interagency Working Group on Sustainable Development Indicators. 1998. *Sustainable Development in the United States: An Experimental Set of Indicators*, Washington, D.C.: US Interagency Working Group on Sustainable Development Indicators.

US Military Academy. 1957. Bugle Notes. West Point, NY: US Military Academy.

US Military Academy. 1999. Available from www.usma.edu; INTERNET.

US National Research Council. 1982. Risk and Decision-Making: Perspective and Research. Washington, D.C.

Vesiland, P. Arne. 1995. Evolution of the American Society of Civil Engineers code of ethics. *Journal of Professional Issues in Engineering Education and Practice.* 121: 4–10.

Voland, G. 1999. *Engineering by Design.* Reading, Mass: Addison-Wesley.

Wade, Jeffry S., John C. Tucker, and Richard G. Hamann. 1993. Comparative analysis of the Florida Everglades and the South American Pantanal. Interamerican Dialogue on Water Management. October 27–30. Miami.

Wales, Charles E., Anne H. Nardi, and Robert A. Stager. 1986. *Professional Decision-Making.* Morgantown: Center for Guided Design, West Virginia University.

Watson, Garth. 1988. *The Civils: The Story of the Institution of Civil Engineers.* London: Thomas Telford.

Wisely, William H. 1974. *The American Civil Engineer: 1852-1974.* New York: American Society of Civil Engineers.

Worker can't pin injury on trailer. 2000. *ENR.* 244:22.

World Book Millennium 2000. 1999. IBM Corporation.

World Commission on Environment and Development. 1987. *From One Earth to One World.* New York: Oxford University Press.

World Commission on Environmental Development. 1987. *Our Common Future.* Oxford, England: World Commission on Environmental Development.

World Future Society. 2001. Available from www.wfs.org; INTERNET.

Wright, Kenneth R., Alfredo Valencia Zegarra, and William L. Lorah. 1999. Ancient Machu Picchu Drainage Engineering. *Journal of Irrigation and Drainage Engineering.* 125:360–369.

Zemke, Ron, Claire Raines, and Bob Filipczak. 2000. *Generations at Work: Managing the Clash of Veterans, Boomers, Xers, and Nexters in Your Workplace.* New York: American Management Association.

# Index

ABET. *See* Accreditation Board of Engineering and Technology
accounting  212–213
accreditation  77
Accreditation Board of Engineering and Technology (ABET)  7, 77
administrative law  223
Age of Iron  53
Age of Rationalism  19
Age of Reason  19
agricultural drainage  55
agriculture  51
American Consulting Engineering Council  38
American Council for Construction Education  128
American Institute of Architects  38
American Institute of Consulting Engineers  38
American Public Works Association  37
American Society for Public Administration  119
American Society of Civil Engineers (ASCE), charter  20; divisions  38, 68, 69; strategic plan  237; code of ethics  243
American Society of Municipal Engineers  37
American Society of Municipal Improvements  37
analysis tools  97–102
antitrust laws  42
Appian Way  16
aqueducts  15, 17, 47
arches  19
ASCE. *See* American Society of Civil Engineers
Association for Standardizating Paving Specifications  37
Association of Consulting Engineers (Britain)  38
Aswan High Dam  51, 53–54
attorneys, working with  229
automobiles. *See* cars
aviation  27

banking  193
Baths of Caracalla, Rome  17
Bechtel, Warren A.  39
Bechtel Corporation  39–40
Bethlehem, Pennsylvania  27
Big Cypress Basin  55
Big Dig, Boston  25, 32
Big Thompson Flood Disaster, 1976  41
Billington, David  53
biotechnology  4

Brindley, James 15, 19
budgeting 185, 207–210; elements of 208
buildings 23–24
built environment 56; law 228
Bureau of Reclamation 30, 51
business environment 4
business law 225
business meetings 126–127
business organization 120
business trends 5
Byzantine Empire 17

California 51
California Department of Transportation (CALTRANS) 33–34
California Water Project 48
CALTRANS. See California Department of Transportation
canals 20, 47, 49, 59
Canon of Ethics 236
capitalism and socialism, compared 188–189
career trends 4
cars, availability of 21, 25
Carter, Jimmy 20, 134
Central America 46
Central and Southern Florida Flood Control District 55
ceremonial structures 15
channelization 55
Chernobyl nuclear plant 31
Chicago 52; fire of 1871 31
cholera outbreaks 47
Chunnel 49
citizenship responsibilities 166–167
city management 37
civics 165–167; civil engineering contributions 168
civil engineering, achievements 66; activities 65; areas of practice 10, 23–31; characteristics 238–239; employment distribution 66–67; origin of term 14–15; phases 13–14; requirements and regulation 77–78
civil engineering education. See education

civil engineering projects 84; examples 75–76
civil engineers, functions 73; management activities 73, 74
civil law 223
Clean Air Act 22
Clean Water Act 22, 57
Clinton, Dewitt 20, 39
Clough, Wayne 4
code of ethics 41, 236, 243
Code of Hammurabi 41
codes and standards 77–78
Cold War 14
Coliseum 16
Colorado 41, 51
combined sewers 27
communication, purposes of 152–155
communication skills 11, 151–152
communications 120–121; advanced 4
communications process 150–151
community planning 46
community service 136
competitive negotiation 104–105
computation 11
computing, advances 98
computing tools 97–102
conceptual design 85
condition assessment 94
conflict resolution 11
congressional committees 178
Constantinople 17
construction 23, 68, 88–90, 127–128
construction documents 108
construction firms 39–40
construction law 225
consulting engineers 38–39, 66, 67, 103–105, 110, 114
Consulting Engineer's Council of the United States 38
contract law 224–225
coordination of operations 121
corrective maintenance 93, 94
cost control 212
Council for Higher Education Accreditation 77
criminal law 224
critical thinking 11; defined 141

Croton Aqueduct 28, 39
da Vinci, Leonardo 17

dams 30; environmental impact 53
Dark Ages 17
decision making 115, 121–123, 175–176; elements of 122
decision support systems 123, 124
decommissioning 95–96
Deming principles 131–132
demolition 95–96
Denver International Airport 25
Department of Justice 42
design 10–11, 86–88; organization and oversight 96; procedures 102–103
design codes 102–103
design professional 106
design review 87–88; best practices 89
disaster management 31
disaster mitigation 46–47
drainage 17
drought 55
Drucker, Peter 137–138
Dust Bowl 46

Eads' bridge 27
early civiliation 13–14, 15–19
earthquakes 31, 46
École Polytechnique 19
economic development 49
economic flows 190
economic impacts 57
economic issues 200
economic trends 192
economics 11; basic principles 188; government role 191–192
education 8–9, 78–81,128, 165–166, 238; curricula 8, 9–10, 79–81; early 36–37; regulation 77; skill levels 80
educators' role 1–2
Egypt 15, 51, 53
Eiffel, Gustave 51–52
Eiffel Tower 45, 51–52, 53
electronic commerce 4
emergency management 31, 57
Empire State Building 52
employers 64–66

employment 194
Endangered Species Act 22, 57
energy development 48–49, 50; environmental impact 56
energy law 228
Engineeers Joint Council 42, 236
Engineer's Council for Professional Development (ECPD) code 42, 236
engineering achievements 65–66
engineering activities, regulation 77–78
engineering artistry 53
engineering careers, trends 63–64
engineering designs, regulation 77–78
engineering economics 204
engineering firms 39
engineering societies. *See also* professional societies
engineering specialty areas 72
engineering work, characteristics 62–63, 73, 74–75
engineering work, types 63–64, 65, 73
entrepreneurship 136
Environment and Water Resources Institute 38
environmental decision tools 204
environmental engineering 27–30, 37, 69–70
environmental impacts 54–55
environmental law 228–229
environmental legislation 22
environmental regulation 57
environmental resources, allocation 201; natural, human and built systems, interaction 202
Erie Canal 20, 25, 49
ethics 11, 41–42, 78, 240
evaporation loss 54
Everglades 33, 54–55
Everglades Forever Act 55
expert witness 229
expositions 45, 52

facilities management 132–133
failures 40–41, 105, 160–161; defining 103

family law 227
federal government 34–36, 67, 170–171; budget process 209
Ferris Wheel 45, 52, 53
final design 87
finance 12, 206
financial control 213–214
financial management 206–207
financial planning 207–210; elements of 208
financing 181
fire protection 47–48
flood control 55
Flood Control Act of 1917 30
Flood Control Act of 1944 22
Florida 33, 54–55
forensic engineering 31
formulas, development of 31
fossil fuels 48
Fourier, Jean Baptiste Joseph 19

Galveston hurricane 31
geotechnical engineering 70–71
Glenwood Canyon 97
goals 114–115, 135
government agencies 167; organization 170; relationships between 170
government branches 174; roles 169–170
government, expenditures 177
government incentives 56–57
government regulation 180–181, 190–191; categories 181
government services 191. See also public services
government trends 5
Grand Coulee Dam 30
Great Basin 50
Great Depression 14, 21
Great Wall of China 17
Greece, ancient 16
Greek city-state 15
Greek culture 17
Guri Dam, Venezuela 30

Hagia Sophia 17
hand construction 59
heritage 11

highways 33–34. See also interstate highways
Hong Kong Airport 25
Honshu Shikoku Bridge system 25, 50
Hoosac Tunnel 26
Hoover Dam 30
Hoover, Herbert 20
House Transportation and Infrastructure Committee 34–36
human resources management 124–126
Hurricane Mitch 46
hurricanes 31, 46, 55
hydraulic engineering 47
hydropower 48

Incas 17
Indus valley 17
Industrial Age 14
Industrial revolution 20, 21
industry-occupation matrix 66, 67, 68
inflation 193
information presentation 101
information technology 4
Information Age 14, 23
infrastructure 49, 176; definition 198; economics of 195–198; environmental impact 195; issues 196; law 227
infrastructure life cycles 84, 85, 105
infrastructure management 96–97
infrastructure sectors 191
infrastructure systems 32, 195; primary 2, 3
infrastructure systems matrix 198–199
Institution of Civil Engineers 19–20
interest groups 183–185
international economic issues 194–195
International Association of Public Works Officials 37
International Association of Street and Sanitation Officials 37
International Federation of Consulting Engineers 38
Internet-based technologies 6

interstate highway system  25, 26, 49, 54, 97
inventory  94
investments  57
irrigation canals  72
irrigation systems  50
Isler, Heinz  53
Istanbul  17

Japan  25, 50
Jubail Industrial City, Saudi Arabia  40
judicial system  222–223
Justinian code  17

Kansas City Regency failure  160
King Louis XV  19
King Minos, Greece  17
Kissimmee River  29, 55
Kissimmee-Okeechobee-Everglades ecosystem  55
Kosciuszko, Thaddeus  36

labor economics  194
Lake Nasser  54
Lake Okeechobee  55
Land Grant Act in 1862  37
Las Vegas  48, 51
law  11; impact on civil engineering  218
leadership  112–113
League of American Municipalities  37
legal issues  125; civil engineering projects
legal methodologies  220–221
legal profession  221
Lewis and Clark  20, 50, 51
local government  33, 67–68, 172–175
Lockwood, Greene Engineers Inc.  39
Loma Prieta earthquake  46
London  48
Los Angeles  48
Louisiana Purchase  20, 50

Machu Picchu  17
Maillart, Robert  53
maintenance  92–94, 130

maintenance management systems  94
Malaysia  52
management  12, 31–32; defining  110–111
management audits  132
management skills  133–135
management strategies  4
management theory  111–112
marketing  133
Maslow, Abraham  112
Maslow's hierarchy of human needs  112, 133–134
materials  4
mathematic models  98–99
Mediterranean River  54
Mesopotamia  15
Middle Ages  42, 48
military engineering  19
minorities  81–82
Mississippi River  21, 27; floods  46
Missouri River Basin  22
models  98–101
monetary system  193
moon settlement  43
Muir, John  29
Mullholland Dam  41
multicriterion decision analysis (MCDA) model  99
municipal water supply system  47

nanotechnology  4
Napoleon  19
national development  177
National Environmental Policy Act  22
National Industrial Recovery Act of 1933  21
National Society for Professional Engineers  38
networking  136
new technologies  4
New Age of Steel and Concrete  53
New Millennium Colloquium  4
New York City  20, 28, 47, 52
New York State  20
Nile River  51, 53
Nineveh  48
Northridge earthquake  46
nuclear energy  48

office buildings 75–76
Okeechobee Basin 55
Online Ethics Center for Engineering and Science 236
operations 91–92; management 128–131
organization 96–97; structure of 117–118; basic levels 118
organizational theory 117–119
owner 107

palaces 15
Panama Canal 49
Pantheon 16
Parthenon 16
performance measurement 130
personnel management. *See* human resources management
Peru 17
Petronas Towers 52
Philadelphia 28, 47
Philadelphia-Lancaster Turnpike 24
Phoenix 48
Pick-Sloan plan 22
Pike, Zebulon 50
pipeline work 93
planning 84–86, 113–116, 176
policy analysis 116
policy issues 178–179
political science 169–170
politics 183–185
population increases 4
Powell, John Wesley 51
preliminary design 86
President's Council on Sustainable Development 202–203
preventative maintenance 93
private finance 214–215
problem solving 11, 144–145
production system 113
productivity 129–130
profession, definition 235–236
professional development 135–136, 147
professional practice 11; laws governing 230–231; regulation 78
professional registration 38
professional societies 19, 38, 81, 114, 237. *See also* specific society name

professional status 233–235
program assessment 132
project financing 192
project management 90–91, 127–128
project planning 139–140
project regulation 180–181
project team 106
property law 227
public administration 32, 119–120
public education 57
public finance 214–215
public funds 27
public health 20, 27–30, 47
public involvement 157–158, 181–183
public policy 178–180
public presentations 160
public relations 133
public sector 192–193
public sector management 119–120
public works 20–21, 37, 67, 96–97, 127–128, 131, 164; engineers' role 6, 7. *See also* Government services
public works managers 114
Public Works Week 7
Pyramids, Egypt 15, 16

quality control 92, 105–108, 131
*Quality in the Constructed Project* 105–108
quantitative methods 32

railroads 20, 21, 25–27, 49
Rational Method 31
regulation, government 12
regulatory law 229–230
rehabilitation 95
Renaissance 18
renewable energy 48
replacement 95
request for proposals 105
resource allocation 195–197
revenue management 210–212
risk 102
risk management 31, 116–117
Robling, John A. 38, 39
Roman Empire 16, 17–18, 47

safety engineering 31
San Francisco earthquake, 1906 31

San Francisco-Oakland Bay Bridge 25, 40
San Francisquito River Basin 41
sand filters 27
Santa Clara River basins 41
Sao Paulo, Brazil 52
School of Bridges and Highways, France 19
Sears Tower 52
SEI 78
sewers, brick-lined 48
SFWMD 33, 54–55
shelter 46
simulation models 98–99
Smeaton, John 15
social changes 2, 239
societal challenges 4
South Florida Water Management District. See SFWMD
southwest United States 48
Soviet Union 53
space exploration 49
St. Francis Dam failure 41
standard procedures 102–103
state government 33–34, 67–68, 171–172
statistical tools 99
Statue of Liberty 45
statutes 191
steam engines 20
Stonehenge 15
storm drainage system, maintenance 92–93
storm drains 27
strategic planning 124
structural art 53
structural models 98–99
structures 23–24
subsidies 215
Suez Canal 21, 25
Suez War 51, 53
supply and demand 189–190
sustainable development 11, 57–58, 202–204; indicators 203, 205

tall buildings 52–53
Taylor, Frederick W. 32
teamwork 106, 126
technical industries 65–66

technical subjects, applied 9–10
technical subjects, basic 9
telecommunication law 227–228
temples 15
Tennessee Valley Authority (TVA) 21, 30
Tennessee-Tombigbee canal 49
Teton Dam failure 116
Thames River 48
Thayer, Sylvanus 36
tort law 226
trans-Atlantic cables 21
transportation engineering 24–27, 71
transportation law 227
trends 3–6, 239
Trump, Donald 52

uncertainty 102
urban arterial streets 76
urban sprawl 51
*Urban Public Works Administration* 96
US Bureau of Reclamation 48
US Congress 178–179
US Department of Education 77
US Military Academy at West Point 20, 36–37
Utah 50

verbal communication 159–160
volunteer organizations 136–137

Washington, George 20, 36
waste management, environmental impact 56
wastewater management 48
wastewater systems 27–30
water law 228–229
water planning 32
water resources engineering 30, 38, 47–48, 50–51, 71–72
water supply systems 17, 27
water treatment plants 76
Water Resources Planning Act 22
Water Resources Planning and Management Division, ASCE 38
waterborne disease 28, 47
watersheds 197
waterways 49

West Point. *See* US Military Academy at West Point
westward expansion 21
wetlands 55
Whiskey Rebellion 24
women 81–82

world fairs 45, 52
World Futures Society 42
World Trade Center 52
World War I 21
World War II 21
written communications 158–159